PRIMER OF MEDICAL RADIOBIOLOGY

Primer of
MEDICAL
RADIOBIOLOGY

ELIZABETH LATORRE TRAVIS
Associate
Department of Radiology
Division of Radiologic Sciences
Medical University of South Carolina
Charleston, South Carolina

YEAR BOOK MEDICAL PUBLISHERS, INC.
35 East Wacker Drive / Chicago

To My Parents

Reprinted, April 1979

Library of Congress Catalog Card Number: 75-11105

International Standard Book Number: 0-8151-8836-6

Preface

In recent years the discipline of radiation biology has become a necessary part of medical and allied health school curricula, providing the basis for the many and varied uses of ionizing radiation in the medical profession. As a result, professionals, besides physicians in the three clinical radiology specialties, are expected to and should have a basic knowledge of the biological effects of radiation. This book is intended for just that purpose—to provide a basic background in radiobiology to technologists in diagnostic radiology, nuclear medicine and radiation therapy. Although primarily written with the technologists in mind, the material and depth of coverage should also serve other personnel such as medical and health physicists who are involved in any of the three clinical specialties utilizing ionizing radiation and desire a basic knowledge of biologic events.

Because more and more patients are being referred by an increasing number of physicians to all three radiologic specialties, this book should also be helpful to medical students, residents and physicians in other specialties, providing them the "why's and wherefore's" of the profession. In addition, it can be useful in undergraduate courses in radiation safety, radiobiology and radiation physics.

It cannot be argued that ionizing radiation is a potentially dangerous tool; for this reason, it is imperative that all persons involved with any of the medical applications of ionizing radiation have a basic understanding of its hazardous biologic potential. This knowledge is particularly important in contemporary times as the public is becoming acutely aware of these potential dangers. On the other hand, professionals should not fear this tool, but regard it with a healthy respect. A knowledge of the biologic effects of ionizing radiation will help to promote these attitudes by enhancing the individual's understanding of his chosen profession, thereby enabling him to more effectively function in his field.

I have tried to deal in concepts as well as specifics, with the idea that understanding and knowledge of the basic concepts provides the individual with a greater working knowledge of the topic. Specifics and applications are easily learned if a concept is understood. This book assumes a minimal, not in-depth, knowledge of radiation physics, for the biologic phenomena are a result of physical characteristics of this tool.

Foreword

At last we radiologic technologists have a text composed specifically for us, especially for students and educators in our profession. As students we have sat through lectures by variously qualified instructors who had no textbook geared to aid in learning the basic effect of radiation upon humans. The available books and articles have been prepared for the radiologists, residents in radiology or researchers. Now we have a textbook presented at a level that students may easily digest. The basics are carefully and thoroughly yet simply described, discussed and illustrated in order that, from the beginning of the student's radiologic technology education, a true comprehension may be gained of the potential of a tool for whose use he or she will eventually be responsible.

I believe that I speak for our profession in expressing profound indebtedness to the author of this text. She became a member of the faculty of the Medical University of South Carolina in 1968 as a radiobiologist in the Division of Basic Radiologic Sciences, Department of Radiology. In addition to her participation in research and lecturing to residents in radiology and to student radiologic technologists, she has become nationally known as an instructor, lecturer and seminar panelist. We are particularly appreciative of her interest in technologists and the time she devotes to our profession.

VIRGINIA A. LAMBE, R.T. (AART)
Assistant Professor of Radiologic Technology
College of Allied Health Sciences
Medical University of South Carolina

Acknowledgments

I would like to thank all my colleagues who assisted in the preparation of this text. First, I am deeply indebted to three friends on whom I depended throughout: to Sa Smith who prepared the many revisions and all final illustrations, to Jim Nicholson for his expertise in photomicrography and photography and to Connie Prynne who not only typed the final manuscript and many of the countless drafts but also kept me organized, was invaluable in editing, collected and checked material and references and performed tasks too numerous to mention. Their tireless and special efforts, able assistance and patience are greatly appreciated.

I am particularly grateful to Alene Keen for her unselfish cooperation, loyalty and assistance in typing many of the manuscript revisions, the glossary and the legends and also to Jay Lucas who assisted in the preparation of the glossary by digging definitions out of this text and other sources. Without their help in this cumbersome task, the glossary could not have been included.

A special thanks is due these individuals: Henry Hargrove, Steve Krech, Rosa Lampkin and Vermel Deer, who provided skillful technical assistance in preparing many of the materials used in this text, and Pat Cumming and Jill Seckel, who assisted in photography.

Special gratitude is expressed to Virginia Lambe, a loyal friend for many years and my "public relations person," to whom I owe a great deal. Her critique of the entire manuscript and identification of terms for the glossary were particularly helpful.

Deep appreciation is extended to Jimmy Fenn, our medical physicist and head of my division, who spent hours reading, evaluating and discussing the manuscript with me. His untiring efforts and continual encouragement, given during his few free hours, resulted in a better text than could have been possible without his help. I would also like to thank my other colleagues who gave their time to critically read this manuscript: Dr. Joseph Watson for his assistance with Chapters 2 and 4; Dr. Roberts Rugh for his many recommendations and assistance, so freely given, with Chapter 6; Dr. G. D. Frey and C. J. Klobukowski for their comments on Chapter 9; Drs. Gordon Hennigar and Ronald Gerughty for their critique of the photomicrographs, assistance with captions and Chapter 5; Dr. James Belli for his critique of Chapters 4 and 10; and Fred Parker, R. T. (Director of the Radiologic Technology Program at MUSC), Don Henning, R. T., Joyce Clarkson, R. T., Drs. Harold Pettit, Keene Wallace, Hugh Scruggs, Richard Marks and Jo Ann Simson for their critique of the entire manuscript.

I cannot exclude all of those colleagues who gave permission for illustrations and diagrams reproduced from their works; the library staff at MUSC, particularly Virginia Miller, for assistance; and those friends and family who encouraged and supported me at all times.

I am very grateful to Ken Hoppens and the staff at Year Book for continual assistance and encouragement throughout this endeavor.

Last but not least, I express my sincere appreciation to all my friends in the American Society of Radiologic Technologists, technology students, medical students and residents who have made me acutely aware of the responsibilities of a teacher.

Despite the able assistance of all of the above, I alone accept full responsibility for this text and for any errors or omissions.

ELIZABETH L. TRAVIS

Table of Contents

1/Review of Cell Biology

Living things, whether plants or animals, are made up of organic material called protoplasm. The smallest unit of protoplasm capable of independent existence is the cell. Simple systems may consist of only one cell; however, more complex systems are usually made up of many types of cells differing in size, shape and function.

In such multicellular systems, cells that serve the same function may be grouped together to form a tissue, e.g., blood and nerve. Some tissues are capable of functioning independently with only one cell type; in most instances, however, several types of tissues exist together to form a functioning unit called an organ, e.g., heart, lungs and stomach. Organs whose functions are interrelated are then grouped to form systems such as the cardiovascular, respiratory and gastrointestinal.

Although profound differences do exist among the various types of cells in a multicellular organism, certain basic functional and morphologic (structural) characteristics are common to all cells. This chapter reviews these common characteristics and functions of cells.

Chemical Composition of the Cell

Protoplasm consists of inorganic and organic compounds either dissolved or suspended in water (HOH). Water is the most abundant constituent of protoplasm. In general, protoplasm is 70–85% water; however, the amount of water may vary depending upon cell type. Water serves many functions in the cell and is necessary for life. It is one of the best solvents known; more chemical substances dissolve in water than in most other liquids. Some of the various functions performed by water are a dispersion medium for compounds in the cell and the transport vehicle for substances that are utilized by or eliminated from the cell. In addition, most of the well-established physiologic activities that occur in the cell do so largely in water and, due to its high capacity to absorb and conduct heat, water aids in protecting the cell from drastic temperature changes.

Cell life is virtually impossible without the presence of mineral salts, the inorganic material in the cell. If a cell were placed in water that did not contain salts (i.e., distilled water), it would die. Two examples of salts are sodium (Na) and potassium (K). The concentration

of these two salts, K predominantly inside and Na predominantly outside the cell, is vital in ensuring that cell death does not occur from
swelling or collapsing. Sodium and potassium perform this function
by maintaining the proper proportion of water in the cell (*osmotic
pressure*).

In addition, salts are necessary for the proper functioning of the
cell; for example, muscle cramps are a result of a loss of calcium salts
from the cell. Salts aid in the production of energy in the cell and in
the conduction of an impulse along a nerve.

There are four major classes of organic compounds in the cell:
protein, carbohydrate, nucleic acid and lipid (Table 1-1). Proteins,
constituting approximately 15% of cell content, are the most plentiful
carbon-containing compounds in the cell. Proteins are composed of
simple units joined together to form a long chain complex. These
simple units, because they are individually stable and have specific
characteristics and properties, are called *monomers*. Structures consisting of monomers joined together to form a chain are termed *polymers*; the process by which polymers are joined is *polymerization*.
Another common name for polymers is *macromolecules*; proteins are
macromolecules.

The building blocks of a protein are *amino acids*, of which there
are 22 known variants in living organisms. These amino acids are
linked by a specific type of bond, termed a *peptide bond*. An astounding number of proteins can be derived from the various combinations of these amino acids. A few well-known proteins are insulin,
egg white (albumin), gelatin and hemoglobin.

An important group of proteins in the cell are *enzymes*, a familiar
example of which is papain, an ingredient in commercial meat tenderizer. Enzymes are catalysts, compounds that increase the rate of chemical reactions. The numerous chemical reactions that occur in the cell
must do so at a very rapid rate; therefore, this group of proteins is essential for the proper functioning of the cell.

Proteins are essential not only to the basic functions of the cell
but also are vital building blocks of many acellular tissues of the body
such as hair and nails. The importance of this group of organic compounds to all cell life and functions is referred to in the name "protein," from the Greek "prōtos" meaning to occupy first place.

Another class of organic compounds in the cell is carbohydrate,
making up approximately 1% of the cellular contents. Carbohydrates
are composed of carbon, hydrogen and oxygen and are the primary
source of energy to the cell. This class can be subdivided into three
categories: *monosaccharides*, *disaccharides* and *polysaccharides*.
The first two categories are commonly referred to as sugars, such as

TABLE 1-1.—ORGANIC COMPOUNDS IN THE CELL

NAME	PERCENT	COMPONENTS	EXAMPLES	PRIMARY FUNCTIONS
Protein	15	Amino acids	Insulin Albumin Hemoglobin Enzymes	Basic building blocks of cells and tissues
Carbohydrate	1	Carbon Hydrogen Oxygen	Starch Glycogen Lactose Sucrose	Provide energy necessary to all basic cellular functions
Nucleic acid	1	Sugar Phosphate Nitrogenous base	DNA RNA	Direct cellular information and transmit genetic information between cells and generations Role in protein synthesis
Lipid	2	Differs with various types	Cholesterol Castor oil Steroids (vitamin D, sex hormones, etc.)	Various functions, e.g., store energy, provide protection

table sugar (a disaccharide), and will readily dissolve in water. Polysaccharides are macromolecules since they consist of a monomer, monosaccharide, polymerized to form a chain.

The third major class of organic compounds, nucleic acid, also may exist as macromolecules in the cell. Two nucleic acids in the cell are deoxyribonucleic acid (*DNA*), and ribonucleic acid (*RNA*). At this point, it is important only to realize that DNA and RNA are polymers; their significance and functions will be discussed in detail later in this chapter.

The final class of organic compounds present in the cell is lipid, more commonly known as fat. Lipids are structural components of the cell membrane and are present in all tissues of the body. This class of organic compounds serves a variety of functions in the cell, such as storage of energy, protection of the body against cold and assistance in digestive processes.

Cell Structure

The cell can be divided into two major sections: the *nucleus* and the *cytoplasm*. The nucleus, which is contained within the cytoplasm, is physically separated from it by a membrane, the *nuclear envelope*. This permits the protoplasm of the cell to be accurately described by location as the *nucleoplasm* (within the nucleus) and the *cytoplasm* (outside the nucleus).

If a cell were examined under a light microscope, its structure would be ill-defined. By use of fixation and staining technics, however, it is possible to visualize more clearly structures in both the nucleus and the cytoplasm; these structures are called *organelles*. Although fixation technics kill the cell, they do preserve structural integrity; thus after fixation the cell retains much of the structural organization present in life. It is also possible to selectively visualize organelles in the cell by using stains that will be absorbed preferentially by certain organelles, thus allowing examination of nuclear and cytoplasmic structures in greater detail.

Cytoplasm

The cytoplasm is the site of all metabolic functions in the cell, including both *anabolism* (building up, *synthesis*) and *catabolism* (breaking down) of organic compounds to provide energy and other requirements for life. Under the light microscope, using staining technics, the organelles of the cell appear to be suspended in a medium termed the *ground cytoplasm*, or *matrix*. The organelles in the cytoplasm are membrane-limited structures which compartmentalize the cytoplasm. If it were not for these membrane-bound organelles, the cell would not be able to function in the highly organized manner in which it does. Figure 1-1 represents the structure of a typical cell.

CELL MEMBRANE.—For many years prior to their ability to examine it, biologists were aware of the existence of a membrane surrounding and encasing the cytoplasm of mammalian cells (Fig. 1-1). Indirect observations of the functions and structure of this cell membrane produced some early information and theories concerning its structure. Recent technological advances, e.g., the electron microscope and freeze fracture technics, have made it possible to examine the structure of the cell membrane in greater detail and to determine its constituents; however, the exact configuration remains conjectural. It is known that the cell membrane is predominantly composed of lipids and proteins and appears to be a nonrigid structure.

The main function of the cell membrane is to monitor all exchanges between intracellular (inside the cell) and extracellular (outside the cell) fluid and its contents, thus maintaining the proper physiologic conditions necessary for life. The cell membrane, therefore, is a selectively permeable structure either prohibiting or permitting the passage of substances into and out of the cell. In addition to its permeability, the membrane can conduct an electrical impulse and has a number of enzymatic functions associated with it.

ENDOPLASMIC RETICULUM.—Although described in the late 1800's, very little was known about the *endoplasmic reticulum* (ER, or ergastoplasm) until the 1940's when the electron microscope and

Fig. 1-1.—Diagram of the structure of a typical animal cell with representative organelles labeled.

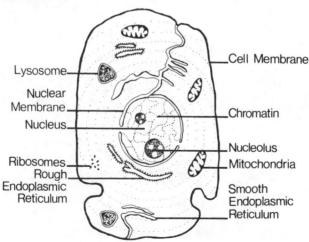

technics in high-speed centrifugation became available. The ER, a double-membrane system, is an irregular network of branching and connecting tubules in the cell that has the capability of remodeling itself in response to changes in intracellular physiologic conditions. Because the space between the two membranes of the ER is observed to be continuous with the space between the layers of the nuclear membrane, this organelle is considered to be an extension of the nuclear membrane, and vice versa.

Two types of endoplasmic reticulum have been observed in cells: granular or rough-surfaced ER and smooth or agranular ER (Fig. 1-2). Rough ER appears granular when stained due to the presence of ribo-

Fig. 1-2.—Electron micrograph of the cytoplasm of a normal rat hepatocyte (liver cell) demonstrating *RER*, rough endoplasmic reticulum; *SER*, smooth endoplasmic reticulum; *M*, mitochondria; and *gly*, glycogen. (Magnification × 37,500.)

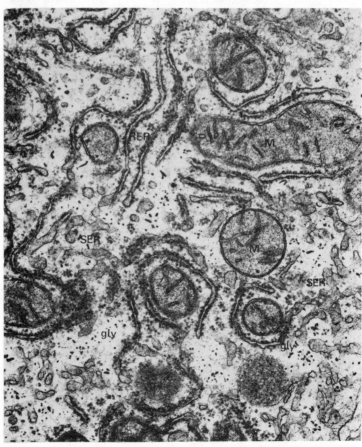

somes on its surface. Ribosomes are not present on the surface of agranular ER, giving it a smooth appearance. Thus the terminology "rough" and "smooth" ER is descriptively appropriate.

The type of endoplasmic reticulum present varies with cell type. In those cells which actively synthesize proteins for export, e.g., pancreatic cells that produce insulin, there is an abundance of granular ER. A reduced amount of granular ER also is present in cells that are synthesizing proteins predominantly for their own use. However, agranular ER is present in cells that are synthesizing products other than proteins. It is well known that granular ER is involved in protein synthesis, but the role of agranular ER is not entirely known—it seems to have a variety of functions depending on cell type.

RIBOSOMES.—Ribosomes are cytoplasmic organelles made up of protein and RNA in approximately equal quantities, either free in the cytoplasm or attached to the endoplasmic reticulum. From the previous discussion of ER, we know that ER with attached ribosomes is actively engaged in protein synthesis; the actual site of protein synthesis in the cell is on the ribosome (see Fig. 1-2).

The synthesis of protein by the cell is a highly organized and efficient procedure involving RNA. Essentially, the process is one in which the nucleus, which controls protein synthesis, sends a message to the cytoplasm indicating which protein is to be made. This message, in the form of a code, is carried by a specific type of RNA appropriately termed "messenger" RNA. The messenger RNA travels to the ribosome, which "reads" the message, translates it and assembles the amino acids in their proper sequence to form the specific protein. At the end of the process, the protein is released from the ribosome. An interesting fact about this organelle is that, regardless of the biologic system studied, ribosomes are similar in size, shape and structure.

MITOCHONDRIA.—These organelles can be visualized with a light microscope and were first observed in the cell as early as 1890. However, it was not until methods of high-speed centrifugation became available that it was possible to harvest mitochondria in bulk and further examine their structure. Mitochondria are the powerhouses of the cell, producing energy for cellular functions by breaking down (catabolizing) nutrients through a process called oxidation. The major source of energy for the cell is carbohydrate, but lipids also can be used for energy production if necessary.

Mitochondria are elliptical structures limited by a double membrane surrounding a central cavity. The inner membrane is folded inward into the central cavity to form a series of shelf-like structures called *cristae* (see Fig. 1-2). The specific enzymes necessary for the production of energy are located on the cristae. These enzymes are not randomly distributed in the mitochondria but are located in a highly organized manner which facilitates the sequence of catabolic reactions for the production of energy.

The number of mitochondria in any particular cell tends to reflect the energy requirements of that cell. For example, cardiac muscle cells requiring a great deal of energy have a large number of mitochondria; lymphocytes, on the other hand, with low energy requirements have only a few mitochondria.

LYSOSOMES.—These organelles, first recognized as a separate entity in the cell in 1955, are single-membrane limited structures which contain enzymes capable of breaking down proteins, DNA and some carbohydrates. These lysosomal enzymes are capable of digesting, or *lysing*, the cell if they are released. Under normal conditions, the enzymes are safely confined within the sac of the lysosome. Many agents, however, are capable of altering the permeability of the lysosomal membrane, causing release of the enzymes. In fact, this was one of the initial theories concerning the mechanism by which radiation kills cells.

GOLGI COMPLEX.—The Golgi complex may exist in a number of morphologic varieties, the most common of which is a stacked set of double membrane structures and small vesicles (little spheres) enclosing a space that may be continuous with the space between the membranes of the endoplasmic reticulum (Fig. 1-3). Although all of the functions of the Golgi complex have not been defined, experimental evidence suggests that this organelle is involved in a variety of functions including involvement in secretion, as a packaging area for products manufactured by the cell for export, carbohydrate synthesis and the binding of other organic compounds to proteins.

Nucleus

The nucleus, which is contained within the cytoplasm, is physically separated from it by a membrane, the *nuclear envelope*. This is a double membrane, the space between the two membranes being continuous with the space between the membranes of the endoplasmic reticulum. The two membranes of the nuclear envelope fuse at various points to form "pores," which are normally closed by a diaphragm appearing to consist of a dense material (see Fig. 1-3). This diaphragm impedes the free exchange of materials between the nucleus and the cytoplasm; however, selective passage of some molecules from the nucleus to the cytoplasm is permitted, as in the case of nuclear RNA. Much of the RNA in the nucleus is contained within a rounded body called the *nucleolus* which, in some cells, is attached to the nuclear membrane (see Fig. 1-3).

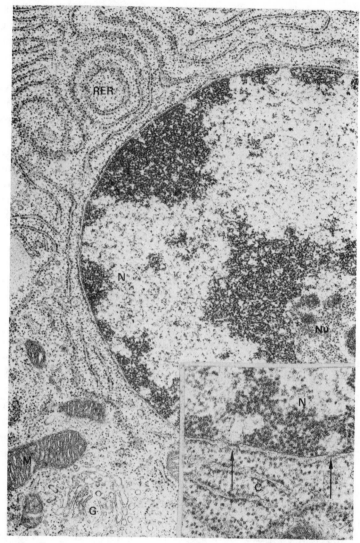

Fig. 1-3.—Electron micrograph of a portion of a nucleus and the cytoplasm of a cell from a rat parotid gland: *N*, nucleus; *Nu*, nucleolus; *RER*, rough endoplasmic reticulum; *G*, golgi; *M*, mitochondria. (Magnification × 25,000.) Note the double membrane surrounding the nucleus except in the region of the pores where it is single (*arrows*). (Magnification × 45,000.) (Courtesy of Dr. J. V. Simson.)

Although the nucleus and the cytoplasm are physically separate, they are mutually dependent upon each other for life. Functionally, the cytoplasm is the site of metabolism, but the nucleus supervises and coordinates cytoplasmic activities. Changes in the nucleus can impair its ability to direct cytoplasmic functions; changes in the cytoplasm will affect the nucleus, since the various functions of the cytoplasm keep the nucleus alive and enable it to reproduce and repair itself.

Contained within the nucleus is the material responsible for directing cytoplasmic activities—the genetic material. This material is contained in *chromosomes*, a word derived from two Latin words meaning "colored bodies." Chromosomes, visible only in dividing cells, appear as dark-stained linear bodies which are constricted at some point by a structure called the *centromere*, a clear region necessary to the movement of the chromosome during cell division. The extensions of the chromosome on either side of the centromere are descriptively termed "arms." The structure of a typical chromosome is given in Figure 1-4.

Chromosomes are the bearers of *genes*, units of genetic material responsible for directing cytoplasmic activity and transmitting hereditary information in the cell. One chromosome contains many genes arranged in a specific linear sequence on the chromosome. Each gene has a specific function to perform in maintaining the life and development of the individual.

Cells in sexually reproducing plants and animals can be generally classified as *germ* cells (*gametes*: female—oocytes; male—spermatozoa) and *somatic* cells (all other cells). These two types of cells differ in the amount of genetic material they contain. In each somatic cell, there are at least two of each kind of gene located on two different chromosomes. These two chromosomes are alike in the sequence and type of genes they carry and are called *homologues*. Therefore, in somatic cells, the chromosomes are paired; each member of a pair is alike, but the pairs themselves are all different.

The number of chromosomes in somatic cells is referred to as the *diploid* or 2n number, which is constant for a given species of plant or animal. However, the 2n number of chromosomes varies with different species; e.g., in humans 2n equals 46, in dogs 2n equals 78, in cats 2n equals 38 and in gorillas and chimpanzees 2n equals 48.

Chromosomes (and therefore genes) in germ cells do not exist in pairs but as individual chromosomes. These individual chromosomes are one of the homologous chromosomes from each of the pairs present in somatic cells. Therefore, the number of chromosomes and genes in germ cells is one-half the number found in somatic cells. This specific number of chromosomes in germ cells is called the *n* or *haploid* number.

Although most somatic cells would die or function abnormally

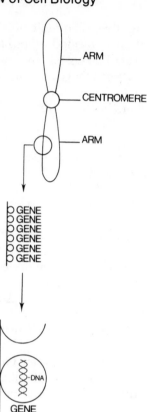

ARM

CENTROMERE

ARM

GENE
GENE
GENE
GENE
GENE
GENE

DNA

GENE

Fig. 1-4.—Diagrammatic representation of the relationship of a chromosome, genes and DNA. **Top,** structure of a typical chromosome. Although shown in the center, the centromere is not located at this site on all chromosomes. **Middle,** enlarged area of the chromosome demonstrating the relationship of genes and chromosomes. **Bottom,** enlargement of one gene, relating DNA to the gene.

with less than the 2*n* number of chromosomes, germ cells can exist with only one-half this number due to their specialization. Metabolically, the germ cells do not require or perform all of the functions of somatic cells; therefore, their limited metabolism makes survival possible with less than the full complement of chromosomes.

Functionally, the germ cells have one specific purpose—they are responsible for reproduction of the species. During fertilization, the male and female gametes unite to form the zygote (fertilized egg). Therefore, by necessity each germ cell must contain only one-half the number of chromosomes to produce a normal zygote containing the essential 2*n* number of chromosomes.

DNA.—The genes are composed of a macromolecule called *deoxyribonucleic acid (DNA)*. The relationship between chromosomes, genes and DNA is diagramed in Figure 1-4. DNA is a double-stranded structure twisted upon itself to form a tightly coiled molecule resembling a spiral staircase. This configuration is termed a *double helix* (Fig. 1-5). The name "deoxyribonucleic acid" suggests the units that make up this macromolecule. The subunits of DNA are a nitrogenous base, a 5-carbon sugar and phosphoric acid. These three components taken together are called a *nucleotide* or, in this case, deoxyribose nucleotide (Table 1-2).

If we un-spiral the DNA macromolecule and consider it as a ladder, we see that the "side rails" of the ladder (the "backbone" of DNA) are composed of sugar molecules joined by a common phosphoric acid molecule (by a *phosphate bond*). The "rungs" of the ladder are made up of the nitrogenous bases (Fig. 1-6). These bases are important constituents of DNA because they are the only units that differ in the macromolecule. The sugar and the phosphoric acid are the same along the DNA strand; however, there are only four kinds of nitrogenous bases present in DNA . These four bases are divided into two categories: the purines, which are adenine and guanine, and the pyrimidines, which are thymine and cytosine (Table 1-3). A base is bonded to a sugar molecule on one or the other side of the ladder. The bases then join the two side rails of the DNA ladder by pairing with each other through hydrogen bonds.

These bases, the purines and the pyrimidines, do not bond randomly with each other but can pair only in a very specific manner: a purine pairs only with a pyrimidine—purines and pyrimidines never

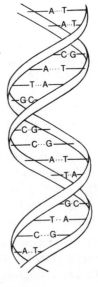

Fig. 1-5.—Diagram of DNA demonstrating double helical structure and the relationship of the four bases to the backbone. *A,* adenine; *T,* thymine; *G,* guanine; *C,* cytosine.

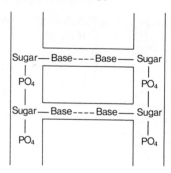

Fig. 1-6.—Ladder-like configuration of an unspiraled DNA molecule; sugar molecules joined by phosphate (PO_4) bonds *(solid line)* form the side rails; nitrogenous bases bonded to each other *(broken line)* and to a sugar molecule on each side rail *(solid line)* form the rungs joining the two backbones.

TABLE 1-2.—CHEMICAL COMPOSITION OF DNA

NUCLEOTIDE
Phosphoric acid
Sugar
Base

TABLE 1-3.—BASES IN DNA

PURINES	PYRIMIDINES
Adenine (A)	Thymine (T)
Guanine (G)	Cytosine (C)

pair with themselves. What is even more fascinating in view of DNA's major role in cell life is that one specific purine can only pair with one specific pyrimidine, i.e., adenine pairs only with thymine and guanine pairs only with cytosine. Because of this characteristic, the two strands of DNA are said to be complementary to each other.

From the above and a large amount of experimental evidence, it has been concluded that DNA is ultimately responsible for directing cellular activity and for transmitting genetic information from cell to cell and from generation to generation. The order that exists in the cell is due to the unique component of the DNA molecule—the nitrogenous base. Because of the uniqueness and specific pairing of the nitrogenous bases, these remarkable constituents determine the genetic information available to the cell, not by themselves but by their linear sequence on the DNA molecule.

This is an astounding concept considering that the various arrangements of only four molecules provide much of the diversity in life as we know it. Even more astonishing is the effect that a single change in one of the bases or in their linear sequence can have on a cell. This alteration of the information carried by DNA, termed a mutation, may be so slight that it will not be grossly manifested by the cell, or it may be a major alteration resulting in cell death. Radiation can cause this change in DNA.

Table 1-4 summarizes the location and function of cellular organelles.

TABLE 1-4.—CELLULAR ORGANELLES

NAME	LOCATION	FUNCTION
Cell membrane	Cytoplasm	Monitors exchanges between cell and environment
Endoplasmic reticulum	Cytoplasm	
Rough ER		Protein synthesis
Smooth ER		Variety of functions in cells making substances other than protein
Ribosomes	Cytoplasm	Protein synthesis
Mitochondria	Cytoplasm	Produce energy by oxidizing carbohydrates and lipids
Lysosomes	Cytoplasm	Contains enzymes capable of destroying the cell
Golgi complex	Cytoplasm	Concentration and segregation of products for secretion; carbohydrate synthesis
Nuclear membrane	Nucleus	Separation of nucleus from cytoplasm; permits selective passage of molecules from nucleus to cytoplasm and vice versa
Nucleolus	Nucleus	Contains RNA
Chromosomes and genes	Nucleus	Control cellular activities

Cell Division

The growth and development of a multicellular organism is dependent on the multiplication of its individual cells. This multiplication process, whereby one cell forms two or more cells, is termed *cell division*. There are two types of cell division: *mitosis*, the method by which somatic cells divide, and *meiosis*, the method by which germ cells divide.

Mitosis

The dynamic physical process by which one "parent" cell divides to form two "daughter" cells is termed mitosis. This process not only results in an approximately equal distribution of all cellular materials between the two daughter cells, but it also maintains the integrity of life of the parent cell by preserving genetic continuity.

Mitosis can be divided into four phases: *prophase, metaphase, anaphase* and *telophase*. Between telophase and prophase of the next mitosis, the cell passes through a nondividing or intermitotic period called *interphase*. Therefore, the four mitotic phases can be considered part of a cycle which begins with interphase and ends with telophase. Before discussing the events which occur during mitosis, it is useful to discuss interphase because it is the activities of the interphase nucleus which make normal mitosis possible.

INTERPHASE.—Microscopically, the interphase nucleus appears nondescript (Fig. 1-7). Using a stain specific for DNA, the genetic material appears clumped in different patterns throughout the nucleoplasm, dependent on cell type and function. However, individual chromosomes are not visible during interphase. Due to this microscopic appearance, interphase was referred to for many years as the "resting" phase of cell division. Improved biochemical technics and the increased use of radionuclides allowed indirect observation and measurement of the cell's activities during interphase. This observation quickly dispelled the idea that this phase is a nondynamic resting period.

As previously stated, somatic cells contain the diploid number of chromosomes and divide by the process of mitosis, defined as the division of nuclear and cytoplasmic material between two daughter

Fig. 1-7.—Photomicrograph of human cervical carcinoma cells illustrating the nucleus (N) with visible chromatin (arrows), nucleolus (Nu) and cytoplasm (C). Cytoplasmic organelles cannot be seen. (H and E stain, magnification × 1000.)

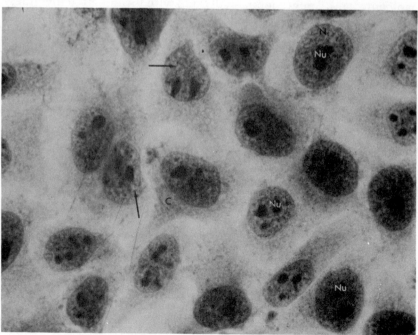

cells, maintaining the integrity of life of the parent cell. Therefore, the daughter cells must have not only the diploid number of chromosomes but also the same genes and therefore an identical sequence of bases on the DNA molecule as the parent cell. At some point in the cell cycle, prior to mitosis, DNA must be duplicated (synthesized) and its integrity maintained. It has been found that DNA synthesis occurs during a portion of the interphase period.

DNA SYNTHESIS.—The process of DNA synthesis doubles the amount of original DNA in such a manner that the new DNA molecules are identical to the original molecule. The two strands of the original molecule each act as a mold or template upon which a new, complementary strand of DNA is synthesized. Initially, the DNA molecule unwinds from its double helical, ladder-like configuration, the bond between two of the bases forming the rungs of the ladder is broken and the two side rails (backbone) of the DNA ladder separate leaving an unpaired purine and pyrimidine base on each strand of the original molecule. Within the cell there is a storehouse of "new" purine and pyrimidine bases which will correctly pair with the unpaired purine and pyrimidine on the two chains of the original DNA molecule. For example, if adenine was on one of the original chains, only thymine will pair with it. On the opposite chain of the original molecule, only a new adenine will bond with the unpaired thymine. A new backbone of sugar and phosphate is simultaneously constructed completing a section of the new chain.

This process gradually continues along each strand of the original DNA molecule until two completely new strands of DNA have been synthesized which are complementary to the original strands (Fig. 1-8). The net result is two DNA molecules that are an exact copy (replica) of the original, providing the necessary complement of DNA to form two daughter cells during mitosis. Therefore, the process of DNA synthesis is often termed *replication* and is designated "S."

It has been found that the process of DNA synthesis does not involve all of interphase. Between telophase and the beginning of DNA synthesis, there is a period called "G_1" (G designating "gap"), when DNA is not replicating. After DNA has been synthesized, there is another period before the cell begins mitosis when DNA again is not replicating; this period is termed "G_2." At the end of G_2 the cell enters the first phase of mitosis—prophase. This cyclic process is diagramed in Figure 1-9. During mitosis, the doubled genetic material becomes visible as chromosomes; each original chromosome actually consists of two identical chromosomes now termed "sister" chromatids. The movement of these bodies through the four phases of mitosis can be observed under the light microscope.

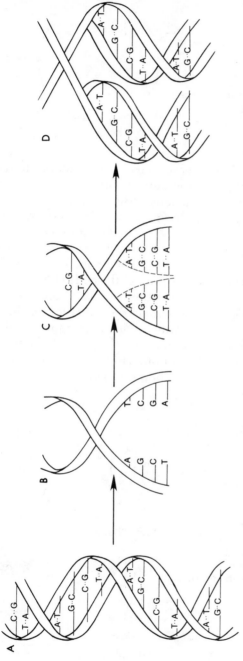

Fig. 1-8.—Schematic representation of DNA replication. **A,** DNA before replication. Note base sequence on molecule. **B,** unwinding and separation of portion of the DNA molecule signals the initiation of DNA replication. Note the unpaired bases on each strand of the molecule. **C,** a storehouse of bases in the cell provides the appropriate partner to join with the unpaired bases on each chain (*dotted lines* indicate site of new backbone). **D,** new backbones are constructed, producing two complete DNA molecules—*both* molecules identical in base composition to the parent. This process continues along the DNA chain until the whole molecule is replicated. *A,* adenine; *T,* thymine; *C,* cytosine; *G,* guanine.

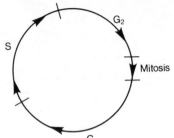

Fig. 1-9.—Diagram of the cell cycle. *M*, mitosis (physical division); *G*$_1$, gap 1; *S*, DNA synthesis; *G*$_2$, gap 2.

PROPHASE.—In early prophase, the genetic material initially appears granular, gradually forming delicate, elongated strands evenly distributed throughout the nucleus (Fig. 1-10A). These delicate strands are the first signs of chromosomal formation. As prophase progresses, the two chromatids of each chromosome shorten and thicken and can be observed moving toward the intact nuclear envelope. During late prophase the nuclear membrane begins to break down, leaving a clear zone in the center of the cell. The chromosomes now move randomly in this zone as they approach the center of the cell (the *equatorial plate*). This signals the initiation of the second phase of mitosis—metaphase. One other important event occurs during prophase, the formation of the *spindle*. The spindle is made up of delicate fibers that extend from one side (pole) of the cell to the other. These fibers are attached at opposite poles of the cell to structures called the *centrioles*.

Fig. 1-10.—Diagram of mitosis in a somatic cell containing four sister chromatids. **A,** prophase—chromatin becomes filamentous and visible. **B,** metaphase—chromatids aligned on the equatorial plate. **C,** anaphase—migration of chromatids along spindle to opposite poles of the cell. **D,** reconstruction of the nuclear membrane and cytokinesis complete the mitotic process.

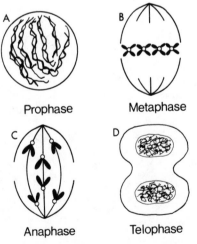

METAPHASE.—Metaphase begins when the chromosomes line up in the center of the cell forming the equatorial plate (Fig. 1-10B). The nuclear membrane is completely broken down at this time, and the chromosomes are free to move around in the cell. The two chromatids of each chromosome now attach to the spindle by means of the centromere and begin to repel each other with their "arms" pointed toward the center of the cell and the centromere pointed toward the pole. Until metaphase, the two chromatids of each chromosome share a common centromere. During metaphase, however, the centromere is duplicated, allowing each chromatid to individually attach to the spindle. The migration of the chromatids along the spindle signals the initiation of anaphase.

ANAPHASE.—During anaphase, the two chromatids repel each other and migrate along the spindle to opposite poles of the cell (Fig. 1-10C). At the completion of anaphase, each duplicated chromosome is located at opposite poles of the cell; this marks the beginning of the last phase of mitosis.

TELOPHASE.—During telophase, the chromosomes lose their definitive appearance and become elongated strands, eventually assuming a homogenous appearance in the nucleus. Simultaneously, the nuclear membrane is reconstructed around the genetic material (Fig. 1-10D). In essence, telophase is the reverse process of prophase.

In addition to nuclear reconstruction, division of the cytoplasm (*cytokinesis*) also occurs during telophase. Indentation of the cell membrane begins to divide the cytoplasm in the vicinity of the equator of the cell. By this process the cytoplasm is divided approximately equally between the two daughter cells. Cytokinesis continues until the cytoplasm is completely separated, forming two daughter cells each having a complete cell membrane.

At the end of telophase the two daughter cells formed are comparable to the original parent cell and contain a full complement of chromosomes ($2n$ set). The cell has now completed mitosis and progresses to interphase. Essentially, the process of mitosis maintains genetic continuity and preserves the status quo. Figure 1-11 consists of photomicrographs of the various stages of mitosis of human cells in tissue culture.

Meiosis

Meiosis is the process by which germ cells divide. This is a specialized type of cell division that reduces the number of chromosomes in the oocyte and spermatozoan from the diploid number ($2n$) to the haploid number (n). As previously stated, a reduction in the chromosome number of germ cells is necessary if the zygote is to contain the

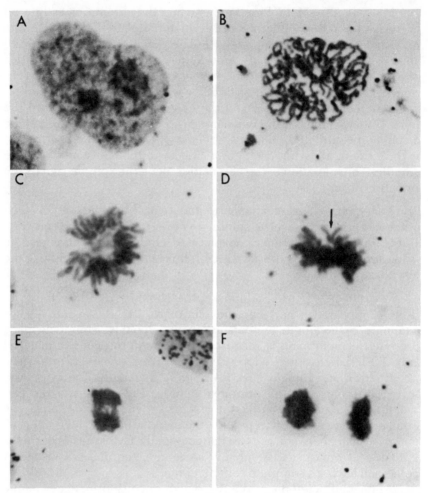

Fig. 1-11.—Photomicrograph of mitosis in a human epithelial cell (HeLa) grown in tissue culture; only the chromosomes are stained. **A,** interphase. **B,** early prophase, note filamentous structure of chromatin. **C,** late prophase, chromosomes approaching equatorial plate. **D,** chromosomes line up forming equatorial plate in a classic view of metaphase. Note visible centromere and arms of one chromosome *(arrow).* **E,** anaphase, chromosomes pulling apart, approaching opposite poles of the cell. **F,** telophase, chromosomes lose their distinct appearance as the cell approaches interphase. (Feulgen stain, magnification × 2000.)

diploid number. If the chromosomes of the germ cells were not reduced from 2n to n (in humans, 46 to 23), the zygote would receive twice the essential number of chromosomes.

Meiosis is a process of reduction division in which the cell di-

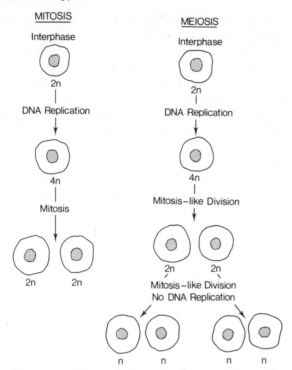

MITOSIS

Interphase

2n

DNA Replication

4n

Mitosis

2n   2n

MEIOSIS

Interphase

2n

DNA Replication

4n

Mitosis–like Division

2n   2n

Mitosis–like Division
No DNA Replication

n   n   n   n

Fig. 1-12.—Diagrammatic comparison of mitosis and meiosis.
Left, mitosis. Process of division in somatic cells. One parent cell with 2n chromosomes forms two daughter cells each with 2n chromosomes, thus maintaining the status quo. Note only one DNA replication process and one division of the cell occurs during mitosis.
Right, meiosis. One parent germ cell with 2n chromosomes forms four daughter cells, each with n chromosomes. Note the first process of meiosis is the same as mitosis, producing two daughter cells with 2n chromosomes; at this point the major difference occurs—germ cells divide again without DNA replication, producing four daughter cells, each with n chromosomes.

vides twice in succession but the chromosomes are duplicated only once. In germ cells, as in somatic cells, DNA synthesis occurs during interphase and the net result is the same—DNA replicates resulting in a duplication of each chromosome forming two chromatids. Therefore, the germ cell begins meiosis with double the amount of genetic material.

The names of the phases of meiosis and mitosis are the same. During meiosis, the movements of germ cell chromosomes are very similar to the movements of somatic cell chromosomes during mitosis. At the end of telophase, the original "parent" germ cell has formed two daughter cells with $2n$ number of chromosomes. The similarities between mitosis and meiosis end at this point.

These daughter cells now undergo a second division of all cellular material, including the chromosomes, without replication of DNA and thus without duplication of the chromosomes. The consequence of these two successive divisions of meiosis is four gametes each containing the haploid number of chromosomes. A diagrammatic comparison of the processes of mitosis and meiosis is given in Figure 1-12.

One other important event occurs during meiosis, which is termed "crossing over." In essence, *crossing over* is the exchange of genes between sister chromatids, allowing for increased genetic variability within a species.

REFERENCES

1. Bloom, W., and Fawcett, D. W.: *A Textbook of Histology* (9th ed.; Philadelphia: W. B. Saunders, 1969).
2. Bresnick, E., and Schwartz, A.: *Functional Dynamics of the Cell* (New York: Academic Press, 1968).
3. Davidson, J. N.: *Biochemistry of Nucleic Acids* (5th ed.; London: Methuen & Company, 1965).
4. DeRobertis, E. D., *et al.*: *Cell Biology* (5th ed.; Philadelphia: W. B. Saunders, 1970).
5. Koehler, J. K.: *Biological Electron Microscopy* (Berlin: Springer-Verlag, 1973).
6. Watson, J. D.: *Molecular Biology of the Gene* (New York: W. A. Benjamin, 1965).
7. White, A., *et al.*: *Principles of Biochemistry* (4th ed.; New York: McGraw-Hill Book Company, 1968).

2/Basic Biologic Interactions of Radiation

The biologic effects of ionizing radiation represent the efforts of living things to deal with energy absorbed by them after an interaction with ionizing radiation. Radiobiology is the study of the sequence of events that follows the absorption of energy from ionizing radiation, the efforts of the organism to compensate for the effects of this energy absorption and the damage to the organism that may be produced.

When discussing the changes that occur in biologic material after an interaction with ionizing radiation, the following generalizations are important to keep in mind:

1. The interaction of radiation in cells is a probability function or a matter of chance, i.e., it may or may not interact and, if interaction occurs, damage may or may not be produced.
2. The initial deposition of energy occurs very rapidly—in a period of approximately 10^{-17} seconds (sec).
3. Radiation interaction in a cell is nonselective—the energy from ionizing radiation is deposited randomly in the cell—no areas of the cell are "chosen" by the radiation.
4. The visible changes in the cells, tissues and organs resulting from an interaction with ionizing radiation are not unique—they cannot be distinguished from damage produced by other types of trauma.
5. The biologic changes resulting from radiation occur only after a period of time (*latent period*) which depends on the initial dose and varies from minutes to weeks or even years.

Basic Interactions of Radiation

An understanding of the visible response of cells, tissues and organisms to ionizing radiation requires an understanding of the initial interactions of radiation in the cell. Extensive research has explored and continues to explore and define the initial changes that lead to the visible response of the cell, tissue or organism.

When ionizing radiation interacts with a cell, ionizations and excitations* are produced in either biologic macromolecules (e.g., DNA) or in the medium in which the cellular organelles are suspended (e.g., water—HOH). Based on the site of these interactions, the action of radiation on the cell can be classified as either direct or indirect.

Direct action occurs when an ionizing particle† interacts with and is absorbed by a biologic macromolecule such as DNA, RNA, protein or enzyme or any other macromolecule in the cell. These ionized macromolecules are now abnormal structures. Thus damage is produced by direct absorption of energy and the subsequent ionization of a biologic macromolecule in the cell.

Compared with direct action, *indirect action* involves absorption of ionizing radiation in the medium in which the molecules are suspended. The molecule in the cell that primarily mediates indirect action is water (HOH).

The absorption of radiation by a water molecule results in the production of an ion pair (HOH^+, HOH^-). This occurs through the following reaction:

$$HOH \xrightarrow{\text{Radiation}} HOH^+ + e^-$$

The free electron (e^-) is captured by another water molecule forming the second ion:

$$HOH + e^- \rightarrow HOH^-$$

The two ions produced by the above reactions are unstable and rapidly dissociate (break down), provided that normal water molecules are present—forming another ion and a free radical†† by the following reactions:

---

*Excitation is a result of radiation interaction in the cell, a possible consequence of which is damage. Once believed to be relatively inefficient for causing damage in molecules, excitation has been shown to have a high efficiency for causing bond breakage and therefore damage in some systems. In general, the effects of excitation on a molecule vary with the complexity and inherent stability of the molecule. The reader should refer to a text on basic radiation physics for further discussion of this phenomenon.

†Although x- and γ-rays are electromagnetic rather than particulate radiations, their interaction in matter involves the transfer of all or part of their energy to an electron of an atom. This electron, a particle, is ejected from the atom and, dependent on its energy, moves rapidly through matter producing numerous ionizations and excitations along its path. Therefore, the primary agent that produces damage from an x- or γ-ray is a *particle*—a so-called fast electron. The reader should refer to a text on basic radiation physics for further explanation.

††Free radicals, symbolized by a dot, e.g., OH· and H·, contain a single unpaired orbital electron that renders them highly reactive because of the tendency of the unpaired electron to pair with another electron.

$$HOH^+ \rightarrow H^+ + OH\cdot$$

$$HOH^- \rightarrow OH^- + H\cdot$$

The ultimate result of the interaction of radiation with water is the formation of an ion pair (H^+, OH^-) and free radicals ($H\cdot$, $OH\cdot$). The consequences of these products to the cell are many and varied. The ion pair may react in one of two ways:

1. The ions can recombine and form a normal water molecule—the net effect in this case will be no damage to the cell—$H^+ + OH^- =$ HOH.
2. The ion pair can chemically react and damage cellular macromolecules.

Generally, because the H^+ and OH^- ions do not contain an excessive amount of energy, the probability that they will recombine, not causing damage in the cell, is great provided they are in the vicinity of each other.

The free radicals produced are extremely reactive due to their chemical and physical properties and can undergo a number of reactions, a few of which are:

1. Recombine with each other producing no damage, e.g., $H\cdot + OH\cdot \rightarrow H_2O$.
2. Join with other free radicals, possibly forming a new molecule that may be damaging to the cell, e.g., $OH\cdot + OH\cdot \rightarrow H_2O_2$ (hydrogen peroxide, an agent toxic to the cell).
3. React with normal molecules and biologic macromolecules in the cell forming new or damaged structures, e.g., $H\cdot + O_2 \rightarrow HO_2$—free radical combined with oxygen forming a new free radical; $RH + H\cdot \rightarrow R\cdot + H_2$—free radical reacts with a biologic molecule (RH) removing H and forming a biologic free radical.

The effects of free radicals in the cell are compounded by their ability to initiate chemical reactions, and therefore damage, at distant sites in the cell. Although many other reactions occur and many other products are formed by the interaction of radiation with water, free radicals are believed to be a major factor in the production of damage in the cell.

In summary, *direct action* produces damage by direct ionization of a biologic macromolecule; *indirect action* produces damage through chemical reactions initiated by the ionization of water. In both cases the primary interaction (ionization) is the same; the definition of direct and indirect action depends on the *site* of ionization and energy absorption in the cell.

An important point to keep in mind is, because there is more water in the cell than any structural component, the probability of radiation damage occurring through indirect action is much greater than the probability of damage occurring through direct action. In addition, indirect action occurs *primarily* but not *exclusively* from free radicals resulting from the ionization of water. The ionization of other cellular constituents, particularly fat, also can result in free radical formation.

Figure 2-1 diagrammatically presents a comparison of direct and indirect action in the cell.

LET and RBE

Linear energy transfer (LET) is a term describing a physical property of a particular type of radiation—the rate at which energy is deposited as a charged particle travels through matter. LET, expressed in keV/microns (μ), or energy deposited per unit distance of path traveled by the particle, is a function of the physical properties of the radiation, i.e., mass and charge. Electromagnetic radiation (x- and γ-rays), although having no mass or charge, produces fast electrons—particles with negligible mass and a charge of minus 1. Because of these physical properties, the probability of an electron interacting with an atom is relatively small; therefore, the interactions of this primary agent of damage are sparse and the ionizations produced are distant from each other. For this reason, x- and γ-rays are termed low LET radiation.

In contrast to electromagnetic radiations, highly ionizing, particulate radiations (e.g., α-particles, neutrons), having appreciable mass and/or charge, have a greater probability of interacting with matter. These types of radiations lose energy rapidly, producing many ionizations in a very short distance. Alpha particles and neutrons are high LET radiations. Some average LET values for different types of radiation are given in Table 2-1 (refer to a text on radiation physics for a more complete discussion of LET).

Fig. 2-1.—Illustration of direct and indirect action of radiation *(wavy arrow)* in the cell. *Shaded areas* enclosing asterisks represent the site of interaction, either in biologic macromolecules (e.g., DNA) or intracellular water (HOH).

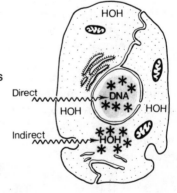

TABLE 2-1.—LET VALUES

TYPE OF RADIATION	LET (KEV/μ)
Cobalt-60	0.3
3 MeV x-ray	0.3
250 keV x-ray	3.0
5 MeV alpha-particle	100.0
Neutron: 19.0 MeV	7.0
2.5 MeV	20.0
Electron: 1.0 MeV	0.25
1.0 keV	12.3

Because of these differences in rate of energy loss, different LET radiations will produce different degrees of the same biologic response, i.e., equal doses of radiations of different LET's do not produce the same biologic response. A term relating the ability of radiations with different LET ranges to produce a specific biologic response is *relative biological effect (RBE)*. RBE is defined as the comparison of a dose of test radiation to a dose of 250 keV x-ray* which produce the same biologic response. This is expressed in the following formula:

$$RBE = \frac{\text{Dose in rads from 250 keV x-ray}}{\begin{array}{c}\text{Dose in rads from another radiation}\\\text{delivered under the same conditions}\end{array}}$$

to produce the *same* biologic effect. Notice the biologic response is the constant, *not* the dose of radiation; what is actually measured is the biologic effectiveness of radiations of different LET's.

Although the relationship between LET and RBE has been discussed in simple terms, it is much more complex. RBE does provide useful information concerning the biologic effectiveness of different types of radiation, but it must be carefully used. RBE is a meaningful value if both the test system used and the biologic endpoint measured are identical. One of the major problems encountered in RBE determinations is that as radiation travels through matter, losing energy, the LET increases after each interaction until the ionizing particle finally stops; therefore, RBE also changes. For this reason, systems and endpoints must be carefully chosen. In addition, other factors such as the chemical environment of the cell will have an effect on the relationship of LET and RBE (discussed in Chapter 4). Another important

---

*250 keV x-ray is the *standard* for determining RBE because of its widespread use at the time the concept of RBE was adopted. Today, although cobalt-60 is more widely used, 250 keV x-ray remains the standard radiation for determining RBE.

point to keep in mind is that the RBE will change depending on the biologic response studied.

Target Theory

As discussed in Chapter 1, the cell consists of many different molecules (enzymes, proteins, RNA, DNA, etc.) and organelles (mitochondria, ribosomes, etc.). All these molecules and organelles have specific functions to perform; however, they all have one common purpose—to work together and keep the cell alive. Each individual structure and molecule in the cell is important, but the number of these molecules in the cell varies. Many copies of some molecules (e.g., enzymes) may be present; not all these molecules may play a role in the life of the cell at all times. Other molecules (e.g., DNA) are present only in the necessary amount, and all are constantly needed for the proper functioning and life of the cell. These latter molecules can be considered *key* molecules in the cell.

The existence of key molecules and their importance in maintaining cellular integrity implies that damage to these molecules can be of far greater consequence to the cell than damage to molecules of which many are present. If more than the necessary number of molecules are present than are needed for normal functioning and a few of these are damaged, one of the "extra" molecules can replace the damaged one and the cell will continue to function normally. However, if a key molecule is damaged, this may present a life-threatening situation to the cell because there is no "extra" molecule present that can continue the function of the damaged molecule.

This concept has led to a theory called *target theory* in terms of radiation damage. When ionizing radiation interacts with one of these key molecules, or within a short distance around one of these key molecules, the name given to this sensitive area is "target." An ionization occurring within the target is called a "hit." The term target must not be interpreted as indicating selective interaction of radiation with a specific cellular site. This is certainly not the case because all ionizing radiation interactions are random events with no specificity or selectivity for the site of interaction. The term target is based solely on the assumption that a random ionization occurring in this area will be of greater consequence to the life of the cell than an ionizing event occurring in another part of the cell.

According to the definitions of target and hit, target theory applies *only* when radiation interacts with the target by direct action. Because living systems contain a great deal of water, indirect action cannot be discounted as causing changes in the target. For this reason, the target theory is not applicable to many living systems. However, this does not negate the concept of key or target molecules in the cell.

Although these targets have eluded positive identification, overwhelming evidence indicates that the nucleus is much more sensitive to radiation damage than the cytoplasm, thus implying the target for radiation is a nuclear constituent. Since DNA is the molecule in the nucleus that controls all cellular activities, logically, an alteration in DNA can be assumed to present more serious consequences to the life of the cell as opposed to an alteration in other cellular constituents such as enzymes or water molecules. Therefore, considering what is known about the functions of molecules and organelles in the cell, DNA is the most likely target for radiation action since an ionization in DNA may present a life-threatening situation to the cell.

Radiation Effects on DNA

Many different types of damage can occur in the DNA molecule as a consequence of radiation. Listed below are only a few types of damage that have been observed after high-dose irradiation of DNA molecules in vitro*:

1. Change or loss of a base.
2. Breakage of the hydrogen bond between the two chains of the DNA molecule.
3. Fracture (break) in the backbone of one chain of the DNA molecule.
4. Fracture in the backbone of both chains of the DNA molecule.
5. Fracture and subsequent cross-linking within the DNA molecule, or from one DNA molecule to another.

The loss or change of a base on the DNA chain results in an alteration of the base sequence (Fig. 2-2A and B). Since it is the sequence of these bases which stores and transmits genetic information, this can be of minor or major consequence to the cell. Regardless of its severity, loss or change of a base is considered a type of *mutation*.

The significance of breakage in the hydrogen bonds between the two strands of DNA is debatable; some investigators consider this an important effect of radiation while others maintain that the tight double-helix configuration of the DNA molecule enhances rapid reformation of the broken bond (Fig. 2-3).

Single-strand breaks in the DNA backbone do occur as a result of radiation (Fig. 2-4). Again, because of the configuration of the DNA molecule and the additional presence of repair enzymes, the break

---

*Literally, in glassware, or in an artificial environment; the opposite of in vivo —in a living system.

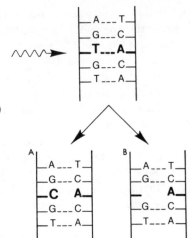

Fig. 2-2.—Change **(A)** or loss **(B)** of a base following radiation *(wavy arrow)* interaction with DNA. Top shows DNA prior to irradiation. **A,** the original base thymine *(T)* has been replaced by cytosine *(C)*. **B,** the original base thymine *(T)* has been lost; none has replaced it. Both these changes alter the base sequence on the DNA strand.

may be repaired with no resultant damage to the cell. A second possibility is the joining of a free radical with the open sites on the broken backbone, thereby preventing proper rejoining, resulting in damage to the cell.

Double breaks in the backbone of the DNA molecule have been observed after irradiation (Fig. 2-5). The number of double-strand breaks is partially dependent on the LET of the radiation. Although both breaks can be repaired, there is a smaller probability of repair occurring in double-strand than in single-strand breaks. If repair does not take place, the DNA chains can separate with greater consequence to the life of the cell.

Cross-linking within the DNA molecule, or between two DNA molecules, is a result of fractures in more than one DNA backbone (Fig. 2-6). If one break occurs in each backbone of the same molecule, or if a break occurs in one backbone of two DNA molecules in close proximity, joining may occur at the sites of these breaks. This results,

Fig. 2-3.—Radiation *(wavy arrow)* interaction with DNA may result in breakage of the hydrogen bond between any two bases, occurring here on the right between thymine *(T)* and adenine *(A)*. DNA prior to irradiation is at the left.

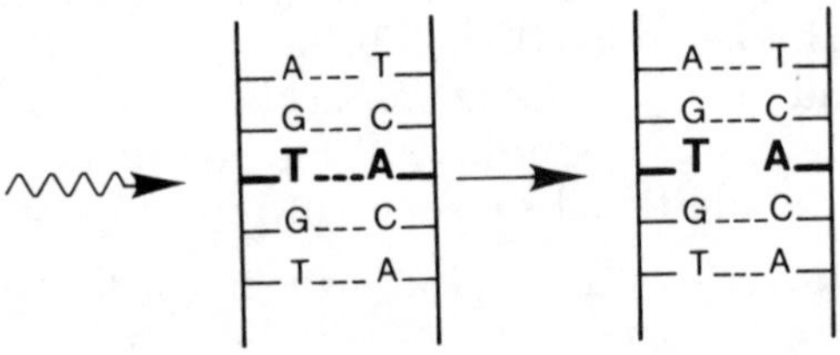

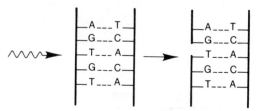

Fig. 2-4.—Irradiation *(wavy arrow)* of DNA *(left)* may result in a single break in the DNA backbone *(right)*.

respectively, in linkages between the two chains of the same DNA molecule, or linkages between two different DNA molecules. Cross-linking is believed to be a result of formation of reactive sites at the points of chain breakage.

It is not within the scope of this text to discuss in greater detail radiation-induced changes in DNA; however, it is important to realize the possible implications to the cell of damage to this structure. Radiation interaction with DNA does not always mean that damage occurs in the cell; much of this damage can be and probably is repaired. But any change in the DNA molecule that is *not* repaired is a mutation resulting in either minor consequences (cell is alive but has some functional impairment) or major consequences (death of the cell). In addition, DNA changes in germ cells may affect future generations, unlike those in somatic cells which affect only the individual.

Radiation Effects on Chromosomes

Although changes in DNA can have serious implications for a cell, also of consequence would be radiation-induced changes in a chromosome resulting in structural changes (breakage) of the chromosome. Admittedly, changes in the DNA molecule are reflected in the chromosome; however, DNA changes are discrete and do not necessarily result in structural chromosome changes.

Some chromosome breaks produced by radiation can be observed

Fig. 2-5.—Another consequence of irradiation of DNA is double breaks in the backbone of both DNA strands *(right)*.

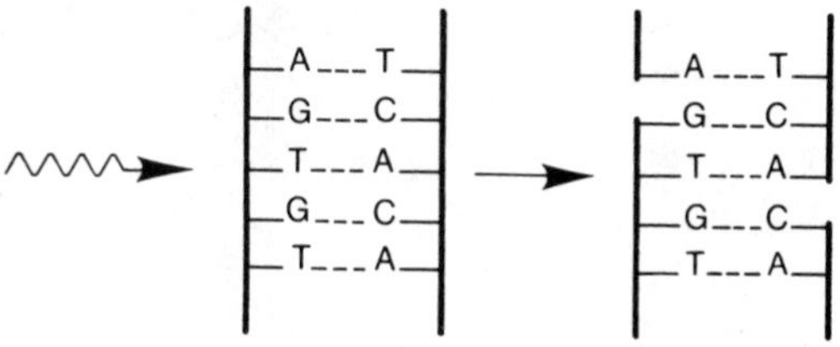

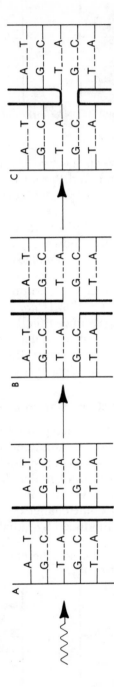

Fig. 2-6.—DNA cross-linking as a result of breaks in two backbones and the formation of reactive sites at these breaks, thus enhancing linkage. Diagrammatically illustrated here. **A,** two DNA strands before irradiation. **B,** DNA that has sustained a break in each backbone as a result of irradiation. **C,** linkage at the site of these breaks.

microscopically during the subsequent postirradiation cell division and are particularly evident during metaphase and anaphase when the chromosomes are shortened and visible. The events that have occurred prior to this time can only be inferred since they are not visible. What are observable are the consequences of these events, i.e., the gross or visible changes in chromosome structure. Radiation-induced chromosome breaks can occur in both somatic cells and in germ cells and can be transmitted during mitosis and meiosis, respectively.

General Chromosome Effects

Structural changes in a chromosome can be produced by either direct ionization of the chromosome as ionizing radiation passes through it (direct action) or by an interaction with the products formed by the ionization of water (indirect action). Regardless of the mechanism of damage, the result is the same—breakage of the chromosome producing two or more chromosomal fragments, each having a broken end. These broken ends have the important capability of being able to join with other broken ends, usually with the first broken end encountered, possibly forming new chromosomes. These new chromosomes may or may not appear structurally different than the chromosome prior to irradiation.

The gross structural changes in a chromosome after irradiation are interchangeably termed *aberrations, lesions* or *anomalies*. A distinction is made between a *chromosome aberration* and a *chromatid aberration*. During the S period of the cell cycle, the chromosome lays down an exact duplicate of itself; these two alike chromosomes are termed *sister chromatids* (see Chapter 1). Chromatid aberrations are those produced in individual chromatids when irradiation occurs *after* DNA synthesis. Because only one chromatid of a pair has been damaged, only one daughter cell will be affected. Chromosome aberrations, on the other hand, are produced when radiation is administered *prior* to DNA synthesis. A resultant break will be replicated if repair is not completed before the initiation of DNA synthesis. In this case, both chromatids will exhibit the break and both daughter cells will inherit a damaged chromatid.

Because of its randomness, radiation may produce a variety of structural changes, some of which are as follows:
1. A single break in one chromosome or chromatid.
2. A single break in separate chromosomes or chromatids.
3. Two or more breaks in the same chromosome or chromatid.
4. "Stickiness," or clumping of the chromosomes.
The general consequence to the cell of these structural changes may be one of the following:

1. Healing with no damage.
2. Loss of a part of the chromosome or chromatid.
3. Rearrangement of the genes (and therefore, the DNA) in the chromosome.

Single Breaks

One Arm of One Chromosome

A single break in one chromosome produces two fragments, one with a centromere and one without a centromere. This can result in two different consequences to the cell. The first possibility, rejoining and healing of the broken ends, has a high probability of occurring due to the close proximity of the chromosome fragments. This process, termed *restitution*, results in no damage to the cell because the chromosome has been restored to its pre-irradiated condition (Fig. 2-7); 95% of single chromosome breaks are believed to heal by restitution.

If irradiation occurred prior to DNA synthesis and restitution has not occurred, both fragments will be duplicated during synthesis, resulting in four chromatids, two with centromeres and two without centromeres, each having a broken end. The two chromatids without a centromere may join forming an *acentric* chromatid, and the two chromatids with centromeres may join forming a *dicentric* chromatid (Fig. 2-8).

The consequences of this type of damage are visible during mitosis (Fig. 2-9). The acentric chromatid will not attach to the mitotic spindle because of the absence of a centromere. Therefore, the amount of genetic information carried on the acentric chromatid will not be transmitted to the daughter cell.

The dicentric chromatid will attach to the spindle; however, the two centromeres will orient the chromatid toward opposite poles of the cell. During anaphase, one centromere will pull the chromatid

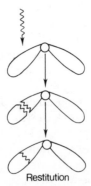

Fig. 2-7.—Radiation interaction with a normal chromosome *(top)* may produce a single break in one chromosome arm *(middle)*, which may be repaired, restoring the chromosome to its pre-irradiated state *(bottom)*. This process is termed restitution.

Restitution

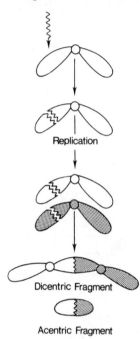

Fig. 2-8.—Formation of an acentric and dicentric fragment resulting from a single break in one arm of a chromosome, which has not been restituted prior to DNA replication.

toward one pole and the other centromere will pull it toward the other pole. As the centromeres continue to migrate to opposite poles of the cell, the chromatid section between them (the junction site of the original broken ends) will be stretched. Eventually, the chromatid will break again, although not necessarily at the point of fusion of the original broken ends. Each chromosome will be drawn into one of the daughter nuclei, and the same process can be repeated since a chromosome with a broken end is still present. If this process continues through many divisions, a major part or even the entire chromosome may eventually be lost from the cell.

The ultimate result of a single chromosome break not repaired before DNA replication is loss of genetic information from both daughter cells, either in the form of a whole chromosome or part of a chromosome, and transmission of incomplete genetic information to both daughter cells.

Whereas a single break in a chromosome before replication will affect both daughter cells, a single break in a chromatid (i.e., after DNA replication) will affect only one daughter cell. The ultimate result in this situation will be the same—loss of genetic information—but the damage will be transmitted to only one daughter cell.

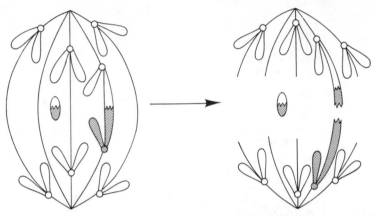

Fig. 2-9.—Diagram of the fate of the dicentric and acentric fragment illustrated in Figure 2-8. The dicentric fragment attaches to the mitotic spindle but migrates to opposite poles of the cell *(left)*, ultimately breaking again *(right)*. The acentric fragment does not attach to the spindle, resulting in loss of genetic information.

One Arm of Two Chromosomes

A single break in one arm of two different chromosomes produces four chromosome fragments—two with centromeres and two without centromeres—resulting in both a greater possible number and variety of aberrations. One aberration formed, a dicentric and acentric aberration, produces the same general consequences to the cell as those outlined in the previous section. However, with this type of break the acentric and dicentric chromosomes are formed by joining of the broken ends of two *different* chromosomes—not as a result of replication of an unrepaired break in *one* chromosome (Fig. 2-10A). For this reason, the consequence to the cell of a single break in two different chromosomes may be more severe because two different chromosomes have sustained damage.

Another possible rearrangement of the fragments is joining of the acentric fragment from one chromosome to the fragment containing the centromere of the other chromosome, and vice versa. This process, termed *translocation*, results in the formation of new, normal-appearing chromosomes, complete with a centromere and the necessary number of genes (Fig. 2-10B). These chromosomes will be transmitted at mitosis or meiosis, and the daughter cells will receive the full complement of genetic information. The only effect will be a change in the order of the genes on the chromosome.

The consequences of this change will be dependent on the cell type in which the translocation occurs. In a somatic cell, a change of

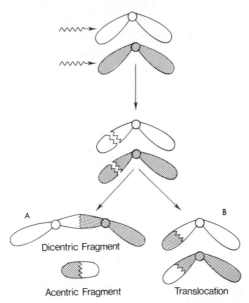

Fig. 2-10.—Two different chromosomes *(top)* may sustain a single break in one arm *(center)* resulting in **A,** formation of a dicentric and acentric fragment, or **B,** translocation of genetic information between the two. In the latter process, two complete chromosomes are formed. Note, however, the exchange of chromosome parts.

gene sequence will not necessarily result in major (life-threatening) consequences to the cell because the necessary number of chromosomes is present. This does not, however, exclude the possibility that the functions these genes control may be altered because of their rearrangement on the chromosome.

Translocation in a germ cell can be of major consequence, particularly if occurring in a gamete that plays a part in conception. The altered chromosome can be transmitted to succeeding generations and, depending on the severity of the damage, may result in either nonviable offspring or offspring which, although surviving into adulthood, may have gross malformations or impaired functions. Of course, the possibility also exists that the effects of the damaged chromosome will never be exhibited in the offspring. The real danger of chromosome translocations is that their effect may be masked and not exhibited for many generations, only to be manifested at a much later time.

Double Breaks

Two or more breaks in a chromosome result in an increased probability of aberrations occurring, with an attendant increase in the probability of damage being exhibited by the cell. More importantly, there is both an increase in type and diversity of aberrations resulting from two or more chromosome breaks. When radiation does cause two breaks in chromosomes, the process of restitution is not as efficient in repairing these breaks, resulting in an overall increase in the number of chromosome aberrations.

The resultant aberrations are dependent on the site of the breaks —two breaks can occur in the same arm of the chromosome, or one break can occur in each arm. The immediate result of a double break is three chromosome fragments—each with a broken end. Only one of these fragments will contain a centromere; the other two will be acentric fragments.

One Arm of One Chromosome

When radiation interacts with and causes two breaks in the same arm of a chromosome, the fragment between the breaks may be deleted. This deleted fragment will not have a centromere, and therefore it will not be transmitted during mitosis. The remaining acentric fragment may join with the fragment having the centromere, re-forming a chromosome with no broken ends but deficient in genetic material (Fig. 2-11A). This chromosome will be transmitted during mitosis, but the daughter cell will inherit a defective chromosome. The consequences of this type of damage will be dependent on the number of genes in the deleted fragment and the functions they controlled.

Deletions are not the sole consequence of two breaks in one arm of a chromosome; a change termed *inversion* can occur. When a chromosome is broken, the fragments can move freely in the cell. Translocations resulting from single breaks are a consequence of the fragments moving throughout the cell, coming in close proximity and joining to other fragments. The fragments produced by double breaks have the same ability to move and become displaced from their original location on the chromosome. The fragment with two broken ends (i.e., the fragment between the two breaks) turns around, thereby reversing its original position on the chromosome, and subsequently rejoins with the fragments comprising the original chromosome (Fig. 2-11B). This process of inversion occurs only when a fragment with two broken ends is present since both are necessary for re-forming a whole chromosome.

In inversion, the re-formed chromosome appears normal and contains all the original genes and the original amount of DNA; only the sequence of genes on the chromosome, thus the sequence of bases on DNA, has been altered. The effects of an inversion on the cell are not a result of a gross structural change but of an alteration of gene and DNA base sequence.

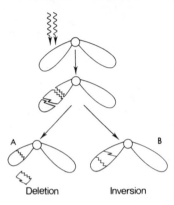

Fig. 2-11.—Two breaks occurring in the same arm of a chromosome (*top* and *middle*) may result in **A,** deletion of the fragment between the two breaks, or **B,** inversion of this fragment, illustrated by a change in the position of the break lines.

The consequences to the cell of an inversion will be dependent on the number of genes and bases in the inverted section and the functions they controlled. The linear sequence of bases on DNA is the ultimate control of all cellular functions; therefore, it is not unrealistic to presume that an inversion can be of major consequence to the cell.

An example might help to clarify the potential damage to a cell of an inversion. Protein synthesis is "ordered" by the linear sequence of bases on DNA, the sequence of bases forming a "code." An inversion changes the order, and therefore the code, carried by the DNA resulting in the production of proteins that are different than the proteins originally coded. These new proteins may or may not be usable by the cell. More importantly, the inversion may order "nonsense" proteins the cell cannot utilize, resulting in decreased amounts of precursor molecules available as building blocks for many substances necessary for life.

Another important point to remember is that different genes on the same chromosome are not necessarily independent of each other. A gene at one end of a chromosome may control the activities of a gene at the other end, which in turn controls the function of a gene elsewhere on the same chromosome. Inversions can upset this interrelationship between genes at different sites on the chromosome and render all of them either nonfunctional or always functional.

Both Arms of One Chromosome

Due to the random nature of radiation, an ionizing event can occur in each arm of the same chromosome. Again, three fragments are formed—two with only one broken end and one with two broken ends. In this case the fragment with two broken ends will contain the centromere, and the fragments with one broken end each will be acentric. Because the fragment with the centromere has two broken ends, this fragment may become inverted before rejoining with the broken ends of the acentric fragments (Fig. 2-12A). The consequence to the cell of this inversion will be dependent on the same previously discussed factors of an inversion of an acentric fragment.

A break in each arm of a chromosome results in very bizarre and unusual chromosome aberrations, an example of which is a ring chromosome. The two broken ends of the fragment with the centromere may move about, coming in close contact. If this occurs, these broken ends follow the rule of all broken ends—they join, forming a ring (Fig. 2-12B). When DNA synthesis occurs, the ring will be replicated and will be transmitted to each daughter cell during anaphase because each chromosome has a centromere.

As the broken ends of the fragment with the centromere are moving about in the cell, they may twist before joining, forming a twisted ring. When DNA synthesis occurs, again, the ring is synthesized; however, in this case the replicated rings are intertwined so that separation at the next anaphase is impossible. One daughter cell will be deficient in one entire chromosome, and the other daughter cell will have one too many chromosomes. In this situation both daughter cells will be seriously affected—as in many other situations, too much of a good thing is as bad as too little!

One other consequence of breaks in each arm of the same chromosome has not yet been discussed: the fate of the two acentric fragments, each with a broken end. As in single breaks, these acentric fragments with their genetic information will be deleted from the cell. If irradiation has occurred after DNA synthesis, only one daughter cell will be affected; if before, both daughter cells will be affected.

Many other chromosome aberrations have been observed as a result of two breaks in chromosomes, including such bizarre configu-

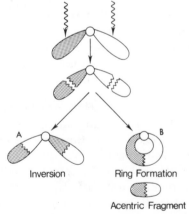

Fig. 2-12.—Radiation interaction with both arms of one chromosome *(top)* may produce a break in each arm *(middle).* The fragment containing the centromere may turn end to end, resulting in **A,** an inversion, or **B,** the arms may join forming a ring. The remaining fragments also join forming an acentric fragment **B.** Note that the inverted chromosome does not appear structurally changed.

Inversion   Ring Formation

Acentric Fragment

rations as star and propeller chromosomes, among others, but the mechanism of these formations is outside the scope of this text. Suffice it to say that the greater the number of chromosomes in a cell that sustain double breaks, the greater the number and the more complex will be the resultant aberrations exhibited by an individual cell. All of the described changes and many others have been observed in the cells of patients undergoing radiation therapy, as well as in individuals occupationally exposed to radiation (Fig. 2-13). Table 2-2 summarizes these chromosome changes.

Chromosome Stickiness

Another phenomenon observed in chromosomes after irradiation, stickiness, occurs in cells already in division at the time of radiation. At metaphase, and particularly during anaphase, the chromosomes appear to be clumped together. Although the mechanism causing this type of damage is still unknown, one possibility is alteration of the chemical composition of the protein component of the chromosome by radiation, allowing the chromosomes to adhere to each other. The chromosomes cannot separate at metaphase and anaphase and form

Fig. 2-13.—Photograph of chromosomes from patients undergoing radiation therapy illustrating many and varied aberrations. (Courtesy of Berdjis, C. C.: *Pathology of Irradiation* [Baltimore: Williams & Wilkins, 1971].)

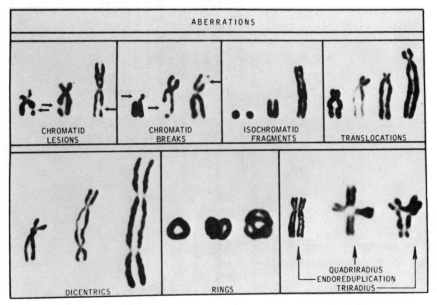

TABLE 2-2.—SUMMARY OF RADIATION-INDUCED CHROMOSOME ABNORMALITIES

TYPE OF DAMAGE	SITE	ABERRATIONS	CONSEQUENCES TO CELL
Single break	One arm of one chromosome	Restitution	No damage
		Acentric and dicentric chromosomes	Loss of genetic information; process repeated with eventual loss of major part or entire chromosome
Single break	One arm of two chromosomes	Acentric and dicentric chromosomes; translocation	Loss of genetic information; process repeated with eventual loss of major part or entire chromosome
Double break	One arm of one chromosome	Deletions Inversions	Change in gene sequence, therefore possible functional alterations
		Inversion	Loss of genetic information; change in gene sequence, therefore possible functional alterations; possible severe cellular effects, e.g., death
Double break	Both arms of one chromosome	Acentric fragments Rings and other unusual and bizarre chromosomes	

bridges between the two opposite poles of the cell. This results in errors in the transmission of genetic information to the daughter cells.

Implications to Humans

Although the implications to humans of DNA and chromosome damage will be discussed thoroughly in a following chapter, it may be of value to put these into perspective at this point. A change in a chromosome means corresponding changes in DNA. Changes in DNA, either an alteration in the amount of DNA resulting from a deletion or an alteration in the sequence of bases on the DNA molecule, result in a change in genetic information in the cell. These changes in genetic information are termed mutations.

It is not always feasible to determine the consequences of mutations to the cell. Some of the consequences will be major, such as death of the cell; however, many mutations are not detectable and do not result in death of the cell. In fact, these discrete changes—difficult if not impossible to detect—can be of great importance to humans and to the general population.

The type of cell (somatic or germ cell) in which the change occurs has important implications. Though mutations in somatic cells result in consequences to that particular individual, they do not have an effect on the general population. Germ cell mutations, on the other hand, do affect the general population because the cell carrying the mutation can play a role in conception and affect future generations.

Factors Affecting Chromosome Damage

Total Dose and Dose Rate

The production of breaks in a chromosome is related to both total radiation dose and dose-rate.* In the case of simple aberrations (i.e., deletions) resulting from single breaks, the number of breaks is directly proportional to the dose, i.e., increasing the dose yields an increasing number of single breaks. However, the number of single breaks is not affected by the dose-rate; a high dose-rate results in the same number of single breaks as when the same total dose is administered slowly. These findings have led to the assumption that chromosome breaks are independent of each other, i.e., a break in a chromosome does not render the chromosome any more or less susceptible to a second break.

---

*Rate at which radiation is delivered; 100 rads may be given in 1 minute (min), therefore the dose-rate is 100 rads/min. If given in 10 min, the dose-rate is 10 radians/min. The latter is a low dose-rate; the former is a high dose-rate.

Complex aberrations (e.g., rings) exhibit a different response to increasing total dose and dose-rate. Complex aberrations are a function of two or more breaks occurring in the same or different chromosomes. Not only must two breaks occur, but both must occur within a short time of each other. Because chromosome breaks are independent of each other and because of the random nature of radiation, as the dose increases there is a greater probability that more chromosomes will sustain single breaks than that the same chromosomes will sustain double breaks. Therefore, an increase in dose does not result in a corresponding linear increase in the number of double breaks and resultant complex aberrations.

Unlike single breaks, double breaks do show a relationship to dose rate. Broken ends of a chromosome remain open for 30 to 60 min.; in addition, any break in a chromosome can heal by restitution. For complex aberrations to occur, two breaks must be open at the same time; therefore, doses administered at a slow rate permit restitution to take place before a second break occurs, decreasing the frequency of complex aberrations. However, if the same total dose is given rapidly, restitution processes are not as efficient in repairing breaks and the probability increases of two breaks remaining open long enough to permit the formation of complex aberrations. For these reasons, low dose-rates result in a decreased number of complex aberrations whereas high dose-rates increase the frequency of complex aberrations.

LET

The type of chromosome aberrations produced by radiation are dependent on LET. Low LET radiations such as x- or γ-rays produce more simple aberrations than complex ones. The probability of an x- or γ-ray inducing two breaks in chromosomes that are in close proximity is very small. Therefore, the number of complex aberrations induced by a given dose of x- or γ-radiation is relatively small. In comparison, high LET radiations such as neutrons or alpha particles have a greater probability of inducing two breaks in the same chromosome at the same time, resulting in a greater number of complex aberrations. It is important to remember that the difference in the number of complex aberrations induced by low LET and high LET radiations is a function of the physical nature of the radiation and not of the chromosome.

Radiation Effects on Other Cellular Constituents

Because of their obvious importance to the life of the cell, much attention has been and continues to be focused on DNA and chromosomes. However, other molecules and organelles in the cell also exhibit radiation effects. Radiation has been observed to cause chain breakage in carbohydrates, structural changes in proteins and alterations in the activity of enzymes. Lipids also exhibit changes after irradiation. Although damage to these molecules does not appear to be as important as damage to DNA, more knowledge is necessary concerning not only the effects of radiation on these molecules but also the implications of this damage to the cell.

Chromosomes are not the only organelles affected by radiation. Permeability of the cell membrane is altered after irradiation, affecting the transport function of the membrane and its ability to keep molecules in or out of the cell. Alteration of membranes by radiation also can affect organelles in the cell that are membrane-bound, e.g., mitochondria and lysosomes. The effect of radiation on cell organelles and the role of damage to these structures in terms of the cellular response to radiation is a subject that requires further study.

REFERENCES

1. Amarose, A. P., *et al.*: Residual chromosome aberrations in female cancer patients after irradiation therapy, Exp. Molec. Path. 7:58, 1967.
2. Alexander, P., and Charles, B. A.: Energy transfer in macromolecules exposed to ionizing radiation, Nature (London) 173:578, 1954.
3. Arnason, T. J., and Morrison, M.: A comparison of the effectiveness of radiations of different energies in producing chromosome breaks, Radiat. Res. 2:91, 1955.
4. Bacq, A. M., and Alexander, P.: *Fundamentals of Radiobiology* (2nd ed.; New York: Pergamon Press, 1961).
5. Bender, M.: Induced aberrations in human chromosomes, Am. J. Pathol. 43:26a, 1963.
6. Berdjis, C. C.: *Pathology of Irradiation* (Baltimore: Williams & Wilkins, 1971).
7. Blois, M. S., Jr.: *Free Radicals in Biological Systems* (New York: Academic Press, 1961).
8. Carlson, J. G.: An analysis of x-ray induced single breaks in neuroblast chromosomes of the grasshopper *(Chortophaga viridifasciata)*, Proc. Natl. Acad. Sci. 27:42, 1941.
9. Casarett, A. P.: *Radiation Biology* (Englewood Cliffs, New Jersey: Prentice-Hall, 1968).
10. Chu, E. H. Y., *et al.*: Types and frequencies of human chromosome aberrations induced by x-rays, Proc. Natl. Acad. Sci. 47:830, 1961.
11. Dewey, W. C., and Humphrey, R. M.: Restitution of radiation induced chromosomal damage in Chinese hamster cells related to the cell's life cycle, Exp. Cell Res. 35:262, 1964.
12. Ebert, M., and Howard, A.: *Radiation Effects in Physics, Chemistry, and Biology* (Amsterdam: North Holland Publishing Company, 1963).
13. Errera, M., and Forssberg, A.: *Mechanisms in Radiobiology* (New York: Academic Press, 1960).

14. Evans, H. J.: Chromosome aberrations induced by ionizing radiation, Int. Rev. Cytol. 13:221, 1962.
15. Giles, N. H., and Tobias, C. A.: Effect of linear energy transfer on radiation induced chromosome aberrations in Tradescantia microspores, Science 120:993, 1954.
16. Kaufman, B. P.: Chromosome Aberrations Induced in Animal Cells by Ionizing Radiations, in Hollander, A. (ed.): Radiation Biology, Vol. 6 (New York: McGraw-Hill Book Company, 1954), p. 627.
17. Lea, D. E.: Actions of Radiation on Living Cells (2nd ed.; Cambridge: Cambridge University Press, 1956).
18. Little, J. B.: Cellular effects of ionizing radiation, N. Engl. J. Med. 278:308, 1968.
19. MacCardle, R. C., and Congdon, T. C.: Mitochondrial changes in hepatic cells in x-irradiated mice, Am. J. Pathol. 31:725, 1955.
20. Moore, R.: Ionizing radiations in chromosomes, J. Coll. Radiol. Australia 9:272, 1965.
21. Müller, H. J.: On the relation between chromosome changes and gene mutations, Brookhaven Symposium on Biology, Vol. 8 (1956), p. 126.
22. Pizzarello, D. J., and Witcofski, R. L.: Basic Radiation Biology (Philadelphia: Lea & Febiger, 1967).
23. Russell, W. L.: The effect of radiation dose rate and fractionation on mutation in mice, in Sobels, F. H. (ed.): Repair from Genetic Damage (New York: Pergamon Press, 1963), p. 205.
24. Sax, K.: Chromosome aberrations induced by x-rays, Genetics 23:494, 1938.
25. Sax, K.: The time factor in x-ray production of chromosome aberrations, Proc. Natl. Acad. Sci. 25:225, 1939.
26. Sax, K.: An analysis of x-ray induced chromosomal aberrations in Tradescantia, Genetics 25:41, 1940.
27. Sax, K.: The effect of ionizing radiation on chromosomes, Q. Rev. Biol. 32:15, 1957.
28. Srb, V.: Immediate and short time changes in cell permeability after x-irradiation, Radiat. Res. 21:308, 1964.
29. Stanton, L.: Basic Medical Radiation Physics (New York: Appleton-Century-Crofts, 1969).
30. Wolff, S., and Luippold, M. E.: Metabolism and chromosome break rejoining, Science 122:231, 1955.
31. Wolff, S.: Radiation Induced Chromosome Aberrations (New York: Columbia University Press, 1963).
32. Wolff, S.: Chromosome aberrations in the cell cycle, Radiat. Res. 33:609, 1968.
33. Zirkle, R. E.: Partial cell irradiation, Adv. Biol. Med. Phys. 5:103, 1957.

3/Radiosensitivity

Soon after Röntgen's discovery, physicians observed that x-rays appeared to destroy the cells of a malignant neoplasm (tumor) without permanently harming the adjacent healthy tissue. This apparently "selective" effect of radiation also was observed among different tissues in healthy animals; some tissues were damaged by doses of radiation that did not appear to harm other tissues.

In 1906 two Frenchmen, J. Bergonié and L. Tribondeau, performed extensive experiments on rodent testicles to further define this observed "selective" effect of radiation. They chose the testes because this organ contains mature cells (spermatozoa), which perform the primary function of the organ, and also contains immature cells (spermatogonia and spermatocytes), which have no function other than to develop into mature, functional cells. Not only do these different populations of cells in the testes vary in function, but their mitotic activity also varies—the immature spermatogonia divide often while the mature spermatogonia never divide.

By observing this one organ, Bergonié and Tribondeau extrapolated their findings to the sensitivity of all cells in the body that have characteristics similar to the cell populations of the testes in terms of mitotic activity and differentiation.[2]

The Law of Bergonié and Tribondeau

After irradiation of the testes, Bergonié and Tribondeau observed that the immature dividing cells were damaged at lower doses than were the mature nondividing cells. Based on these observations of the response of the different cell populations in the testes, they formulated a basic law concerning radiation sensitivity for all cells in the body. In general terms, their law states that ionizing radiation is more effective against cells that are actively mitotic and undifferentiated and have a long dividing future.

From their observations, Bergonié and Tribondeau defined sensitivity in terms of specific characteristics of the cells studied—mitotic activity and differentiation. Sensitivity, defined according to these specific or *inherent* cellular characteristics, is therefore based on characteristics of the cell and *not* on the radiation. Bergonié and Tribondeau's criteria for cellular radiation sensitivity can be interpreted as determinants of the inherent susceptibility of a cell to radiation damage.

In 1925 P. Ancel and P. Vitemberger modified the law of Bergonié and Tribondeau by proposing that the inherent susceptibility of any cell to damage by ionizing radiation is the same but that the time of appearance of radiation-induced damage differs among different types of cells. In a series of extensive experiments on mammalian systems, they concluded that the appearance of radiation damage is influenced by two factors: (1) the biologic stress on the cell and (2) the conditions to which the cell is exposed pre- and postirradiation.[1]

Ancel and Vitemberger postulated that the greatest influence on radiosensitivity is the biologic stress placed on the cell and that the most important biologic stress is the necessity for division. In their terms, all cells will be damaged to the same degree by a given dose of radiation, i.e., all cells are similar in terms of inherent susceptibility, but the damage will be expressed only *if and when* the cell divides. Table 3-1 compares radiation sensitivity as defined by Bergonié and Tribondeau with that defined by Ancel and Vitemberger.

Although Ancel and Vitemberger expressed radiosensitivity in somewhat different terms than did Bergonié and Tribondeau, they still placed the major emphasis on mitotic activity, concurring with the law of Bergonié and Tribondeau. In this text the law of Bergonié and Tribondeau will be generally accepted as a valid expression of cellular radiosensitivity. Although there are exceptions to this law, it still remains a basic and useful guide for determining cellular radiosensitivity.

Differentiation

One aspect of the law of Bergonié and Tribondeau that may need clarification is the term "differentiation." A differentiated cell is one which is specialized functionally and/or morphologically (structurally); it can be considered a mature cell, or *end cell*, in a population. An undifferentiated cell is one that has few specialized morphologic or functional characteristics; it is an immature cell whose primary function is to divide, thus providing cells to maintain its own popula-

TABLE 3-1.—COMPARISON OF RADIOSENSITIVITY
DETERMINANTS

BERGONIÉ AND TRIBONDEAU	ANCEL AND VITEMBERGER
Mitotic activity	Biologic stress, mitotic activity
Differentiation	Pre- and postirradiation conditions
Dividing future	

tion and to replace mature cells lost from the end cell population. Undifferentiated cells can be considered *precursor*, or *stem cells*, in a population.

An example of a tissue that contains a series of cells in various stages of differentiation is the testis. The spermatozoon is the mature, nondividing cell that is morphologically and functionally specialized. However, since mature sperm are periodically lost, more cells must replace them. These cells, also present in the testis, are immature Type A spermatogonia; their principal function is to divide and supply cells that will maturate into spermatozoa. The spermatozoon is a differentiated cell—it is the end cell in the population; the spermatogonium is an undifferentiated cell—the stem cell for the mature spermatozoon. The process by which immature spermatogonia become mature spermatozoa is termed differentiation (Fig. 3-1).

Another example of a differentiated cell is the erythrocyte (red blood cell, or RBC). Just as the spermatozoon is the mature end cell in the testis, the RBC is the mature, differentiated cell in the red cell line of the hemopoietic system. The major function of the RBC is to transport oxygen to cells of the body. Not only is this cell specialized in function, but it also is specialized in structure; the RBC differs from other cells in the body in that it does not have a nucleus. Therefore, both morphologically and functionally, RBC's are differentiated cells. The average lifetime of RBC's in the circulating blood is 120 days, necessitating a continual replacement of these cells by newly produced cells. The stem cell for the RBC, the erythroblast, is present in the bone marrow and is an undifferentiated cell that divides and supplies cells which will differentiate to become erythrocytes.

Cell Populations

An understanding of cellular sensitivity to radiation is dependent on an understanding of the characteristics of various cellular popula-

Fig. 3-1.—Diagrammatic representation of the testes illustrating differentiation. The cell becomes more differentiated as it progresses from spermatogonium (stem cell) to sperm (end cell).

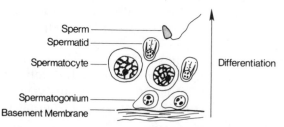

tions. Rubin and Casarett (1968) have defined five basic categories of cell populations (Table 3-2).

Vegetative Intermitotic Cells (VIM)

These are rapidly dividing, undifferentiated cells that have a short lifetime. According to the law of Bergonié and Tribondeau, these cells comprise the most sensitive group of cells in the body. Examples of VIM cells are basal cells of the epidermis, crypt cells of the intestines, Type A spermatogonia and erythroblasts.

Differentiating Intermitotic Cells (DIM)

DIM cells are produced by division of VIM cells and, although actively mitotic, they are more differentiated than VIM cells. Therefore, these cells are less sensitive (or more resistant) to radiation than are the VIM cells. Examples of DIM cells are intermediate and Type B spermatogonia.

Multipotential Connective Tissue Cells

These cells divide irregularly and are more differentiated than either VIM or DIM cells; they are intermediate in sensitivity to radiation. Cells included in this category are endothelial cells (cells lining the blood vessels) and fibroblasts (cells that comprise connective tissue).

Reverting Postmitotic Cells (RPM)

Cells in this category normally do not undergo mitosis; however, they retain the capability of division under specific circumstances. RPM cells are long-lived as individuals and are more differentiated than cells of the previous categories; therefore, these cells are relatively radioresistant. Examples of RPM cells are liver cells and the mature lymphocyte. The mature lymphocyte is included in this category because of its mitotic characteristics—it does not usually divide but has the capability of dividing when a stimulus is present. The lymphocyte also is a differentiated cell; however, in contrast to other RPM cells that are relatively radioresistant, the mature lymphocyte is very radiosensitive. It is one important exception to the general law of Bergonié and Tribondeau.

Fixed Postmitotic Cells (FPM)

FPM cells do not divide. These cells are highly differentiated both morphologically and functionally and therefore are resistant to radiation. In fact, this category comprises the group of cells most resistant to radiation. Some of the cells in this category have long lives, whereas others are relatively short-lived. When the short-lived cells die, they are replaced by differentiating (DIM) cells; other cells in this category may not be replaced if cell death occurs. Examples of cells in this category include some nerve cells, muscle cells, erythrocytes (RBC's) and spermatozoa.

TABLE 3-2.—CHARACTERISTICS AND RADIOSENSITIVITY OF CELL POPULATIONS

CELL TYPE	CHARACTERISTICS	EXAMPLES	RADIOSENSITIVITY
VIM	Rapidly dividing; undifferentiated; do not differentiate between divisions	Type A spermatogonia Erythroblasts Crypt cells of intestines Basal cells of epidermis	Most radiosensitive
DIM	Actively dividing; more differentiated than VIMs; differentiate between divisions	Intermediate spermatogonia Myelocytes	Relatively radiosensitive
Multipotential connective tissue cell	Irregularly dividing; more differentiated than VIMs or DIMs	Endothelial cells Fibroblasts	Intermediate in radiosensitivity
RPM	Do not normally divide but retain capability of division; differentiated	Parenchymal cells of liver Lymphocytes*	Relatively radioresistant
FPM	Do not divide; differentiated	Some nerve cells Muscle cells Erythrocytes (RBCs) Spermatozoa	Most radioresistant

*Lymphocytes, although classified as relatively radioresistant by their characteristics, are very radiosensitive.

Below is a classification of cells according to decreasing radiosensitivity (i.e., radiosensitive to radioresistant) using cell death as the determining factor (Table 3-3).

GROUP 1.—Mature lymphocytes (one of the two major classes of circulating white blood cells), erythroblasts (red blood cell precursors) and certain of the spermatogonia (the most primitive cells in the spermatogenic series) are the most radiosensitive mammalian cells.

GROUP 2.—Granulosa cells (cells surrounding the ovum), myelocytes (cells in the bone marrow that are precursors to circulating white cells), intestinal crypt cells and germinal cells of the epidermis of the skin. This group of cells is less sensitive than is Group 1.

GROUP 3.—Gastric gland cells and endothelial (lining) cells of small blood vessels comprise the next most sensitive group. Cells in this category are only moderately sensitive to radiation. These include osteoblasts (bone-forming cells), osteoclasts (bone-resorbing cells), chondroblasts (precursor cells which form cartilage), spermatocytes and spermatids (cells in the spermatogenic series in the testes).

GROUP 4.—Somewhat more radioresistant are granulocytes (one

TABLE 3-3.—CELL CLASSIFICATION BY DECREASING
SENSITIVITY*

GROUP	SENSITIVITY	EXAMPLES
1	Highly radiosensitive	Mature lymphocytes Erythroblasts Certain spermatogonia
2	Relatively radiosensitive	Granulosa cells Myelocytes Intestinal crypt cells Basal cells of epidermis
3	Intermediate in sensitivity	Endothelial cells Gastric gland cells Osteoblasts Chondroblasts Spermatocytes Spermatids
4	Relatively radioresistant	Granulocytes Osteocytes Spermatozoa Erythrocytes
5	Highly radioresistant	Fibrocytes Chondrocytes Muscle cells Nerve cells

*From Casarett, A. P.: *Radiation Biology* (Englewood Cliffs, New Jersey: Prentice-Hall, 1968).

type of white blood cell), osteocytes (mature bone cells), spermatozoa, superficial cells of the gastrointestinal tract and erythrocytes.

GROUP 5.—The relatively radioresistant cells in this group include fibrocytes (cells of mature connective tissue) and chondrocytes (mature cartilage cells). The least radiosensitive (most radioresistant) cells of the adult mammalian body are mature muscle and nerve cells.

Tissue and Organ Sensitivity

With an understanding of cellular sensitivity to radiation it is now possible to classify tissues and organs in terms of sensitivity. As is to be expected, tissues and organs that contain radiosensitive cells will be sensitive to radiation and, conversely, tissues and organs that contain radioresistant cells will be resistant to radiation.

Tissues and organs are made up of two compartments: the *parenchymal* compartment which contains the cells characteristic of that individual tissue or organ and the *stromal* compartment composed of connective tissue and vasculature which makes up the supporting structure of the organ (Fig. 3-2).

The parenchymal compartment of tissues and organs may be composed of one or more than one category of cells. The testis is an example of an organ that contains more than one category of cells: stem cells—Type A spermatogonia (VIM cells); intermediate cells—Type B spermatogonia, spermatocytes and spermatids (DIM cells); and mature, functional cells—spermatozoa (FPM cells). Another example is the hemopoietic system; the bone marrow contains the undifferentiated stem cells, and the circulating blood contains the mature end cell. Two other examples are the skin and the intestinal tract.

In these types of organs where the parenchymal compartment is composed of various cellular populations, cells flow from the stem cell section to the differentiated section to the end cell section as needed (Fig. 3-3).

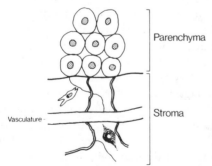

Fig. 3-2.—Diagram of the parenchymal and stromal compartments of any organ. The parenchymal compartment contains cells typical of the organ, whereas the stromal compartment contains connective tissue *(light-colored background),* vasculature and other cells such as mast cells *(below vessel)* and fibrocytes *(above vessel).*

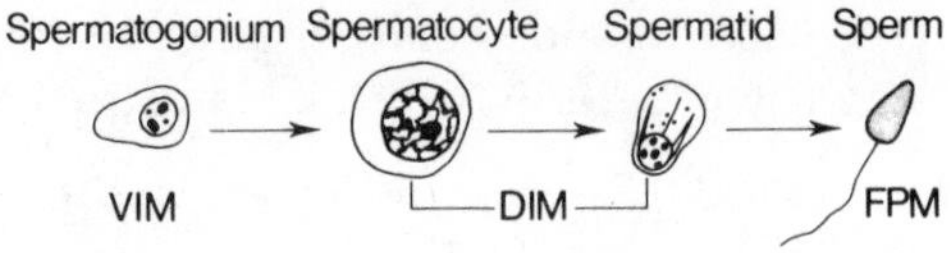

Fig. 3-3.—Flow of cell from the Spermatogonium Spermatocyte Spermatid Sperm
stem cell compartment *(VIM)* to
differentiating compartment
(DIM) to the end cell compartment *(FPM),* illustrated by cells
of the testis.

Examples of tissues and organs whose parenchymal compartments are composed of only RPM cells or FPM cells are the liver,
muscle, brain and spinal cord. The hepatic cells of the liver are RPM
cells and divide only when the need exists. If a partial hepatectomy is
performed, the hepatic cells will begin to divide and replace the part
of the liver that has been removed. However, most cells of the brain
and most muscle cells do not retain the capability of division; therefore, these tissues and organs are composed of FPM cells.

Regardless of the population of cells in the parenchymal compartment, all tissues and organs will have a supporting stromal compartment composed of connective tissue and vasculature (multipotential
connective tissue cells).

The sensitivity of a tissue or organ to radiation is a function of the
most sensitive cell in that particular tissue or organ, i.e., organs that
contain radiosensitive cells will be radiosensitive (e.g., testis and
bone marrow), whereas organs that contain radioresistant cells will be
radioresistant (e.g., liver, muscle and nerve). In a tissue or organ that
contains a series of developing cells (testis, blood, etc.), sensitivity of
the organ will be a function of the most sensitive cell.

Mechanisms of Radiation Damage

The mechanism of radiation damage in radiosensitive and radioresistant organs differs and is a function of the sensitivity of the cell
populations comprising *both* the parenchymal *and* stromal compartments, particularly the vasculature. In tissues and organs that contain
parenchymal cells (VIM and DIM), which are more radiosensitive
than cells comprising the stroma, damage is due to destruction of the
radiosensitive parenchymal cells of that tissue or organ. For example,
sterility occurs because the immature Type A spermatogonia (the
stem cells) in the testis have been destroyed by radiation resulting in a
depletion of mature spermatozoa. Likewise, loss of cells from the circulating blood is usually due to damage to radiosensitive stem cells in
the bone marrow rather than to damage to radioresistant circulating
blood cells. Although damage such as narrowing and occlusion of
blood vessels also may occur in the vasculature of these organs, it is
not the primary contributor to damage of the parenchymal cells (Fig.

3-4). In addition, changes in parenchymal cells occur at lower doses than changes in stromal cells.

However, unlike radiosensitive organs, the cells of the stromal compartment of radioresistant organs are more sensitive to radiation than the parenchymal cells (RPM and FPM) of the organ; therefore, cells of the stroma exhibit changes at lower doses than parenchymal cells. Damage to these organs (e.g., liver, muscle and brain) occurs indirectly through damage to cells of the vascular stroma. This damage may cause narrowing and occlusion of the vessels resulting in a decreased blood supply to the irradiated organ with an attendant loss of nutrients and oxygen, both of which are necessary for the life of the parenchymal cells. Therefore, damage in radioresistant organs is usually caused indirectly through damage to the vasculature. In these organs the vasculature is a major contributor to radiation damage (Fig. 3-5).

It is important to remember, however, that even though mature cells (RPM and FPM) are "resistant" to radiation, they are not "immune" to radiation and *can* be directly damaged by high doses.

Table 3-4 presents a classification of the relative sensitivity of various organs to radiation, based on *hypoplasia* (loss of cells) from the parenchymal compartment at *two months* postirradiation.

Fig. 3-4.—Illustration of the mechanism of damage in a radiosensitive organ (testis). Before irradiation the parenchyma contains VIM, DIM and FPM cells and a connective tissue stroma. After irradiation there is a decrease in all types of parenchymal cells with time, resulting in damage. Regeneration of VIM cells may occur, either partially or completely restoring the organ to its pre-irradiated condition (recovery). Although the vasculature after irradiation becomes tortuous and occluded, it is not the primary contributor to irradiation damage in the organ.

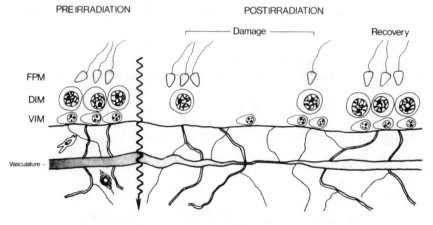

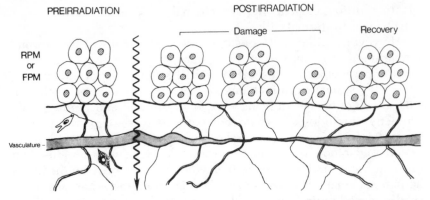

Fig. 3-5.—Illustration of damage in radioresistant organs. Note that in comparison to radiosensitive organs, a decrease in parenchymal cells occurs at a later postirradiation time. Recovery of the parenchymal compartment to its pre-irradiated condition also may take place in this system; damage in these organs occurs indirectly, primarily through vascular damage.

Evaluating Radiosensitivity

There are many biologic endpoints that can be chosen to compare the radiosensitivity of different populations of cells—chromosome breaks and cell death are only two examples. Probably the most widely used endpoint for the determination of cell sensitivity is cell death or morphological changes in the cells that are indicative of cell death. One important point must be remembered when comparing the radiosensitivity of two different populations of cells—regardless of the endpoint used to determine sensitivity, the same endpoint must be used for both populations. For example, if cell death is the criterion chosen to measure the sensitivity of spermatogonia, this same criterion must be used to compare the sensitivity of spermatozoa to spermatogonia.

The criteria used also must be specified when discussing the radiation sensitivity of cell populations. This is necessary because sensitivity may vary with the endpoint chosen. For example, cell A may be more sensitive to radiation in terms of cell death than is cell B, but cell B may exhibit simple chromosomal aberrations at lower doses than does cell A. This is especially true in terms of functional changes. In many instances, cells that are resistant morphologically are sensitive functionally. This topic will be discussed further in another chapter.

TABLE 3-4.—VARIOUS ORGANS IN DECREASING ORDER OF RADIOSENSITIVITY BASED ON HYPOPLASIA IN THE PARENCHYMA OF THE ORGAN TWO MONTHS POSTIRRADIATION*

ORGAN	RADIOSENSITIVITY	CHIEF MECHANISM OF HYPOPLASIA
Lymphoid organs, bone marrow, blood, testes, ovaries, intestines	High	Destruction of radiosensitive parenchymal cells
Skin and other organs with epithelial cell lining (cornea, oral cavity, esophagus, rectum, bladder, vagina, uterine cervix, ureters)	Fairly high	Destruction of radiosensitive parenchymal cells of the epithelial lining
Optic lens, stomach	Medium	Destruction of mitotically active epithelial cells
Growing cartilage		Destruction of mitotically active chondroblasts, plus some damage to the fine vasculature and connective tissues
Fine vasculature		Damage to the endothelium
Growing bone		Destruction of mitotically active chondroblasts or osteoblasts, plus some damage to the fine vasculature
Mature cartilage or bone; salivary glands, respiratory organs, kidneys, liver, pancreas, thyroid, adrenal and pituitary glands	Fairly low	Hypoplasia of parenchymal cells is secondary to damage to the vasculature and connective tissue (minor contribution to hypoplasia from direct effects on parenchymal cells)
Muscle, brain, spinal cord	Low	Hypoplasia of parenchymal cells is secondary to damage to the vasculature and connective tissue (minor contribution to hypoplasia from direct effects on parenchymal cells)

*From Rubin, P., and Casarett, G. W.: Clinical Radiation Pathology (Philadelphia: W. B. Saunders, 1968).

Inherent versus Conditional Sensitivity

Bergonié and Tribondeau's law stressing mitotic activity and differentiation as determinants of the inherent sensitivity of a cell to radiation is generally acceptable; however, the cellular response to radiation can be modified by external factors such as chemicals and the LET of the radiation. This modification of radiation response by external factors can be defined as *conditional sensitivity*. The inherent susceptibility of the cell to radiation damage is not changed, but conditions to which the cell is exposed either pre-irradiation, during irradiation or postirradiation are changed. These conditions result in an apparent increase or decrease in the radiosensitivity of the cell.

As defined in this text, conditional sensitivity is equivalent to the second factor influencing the time of manifestation of radiation damage, as proposed by Ancel and Vitemberger.

REFERENCES

1. Ancel, P., and Vitemberger, P.: Sur la radiosensibilité cellulaire, C. R. Soc. Biol. 92:517, 1925.
2. Bergonié, J., and Tribondeau, L.: De quelques résultats de la radiothérapie et essai de fixation d'une technique rationelle, C. R. Acad. Sci. (Paris) 143:983, 1906.
3. Casarett, A. P.: *Radiation Biology* (Englewood Cliffs, New Jersey: Prentice-Hall, 1968).
4. Rubin, P., and Casarett, G. W.: *Clinical Radiation Pathology*, Vols. I and II (Philadelphia: W. B. Saunders, 1968).

4/Cellular Response to Radiation

As early as eleven years after the discovery of radiation, radiation injuries were reported in occupationally exposed persons. People, fascinated with these "magic" rays which allowed them to see inside the living body and unaware of the dangers involved, used radiation indiscriminately. One of the first reported cases of radiation damage involved a physician who had lost his hair because he had allowed another physician to repeatedly use x-rays to "see" inside his skull! These reports of radiation injuries led individuals to observe and investigate the biologic effects of radiation. By studying responses such as skin erythema and hair loss, they proposed basic postulates concerning the response of different cell populations to radiation. It was not until the 1920's, however, that technics were developed to study individual cells and their response to radiation. The ingenuity and imagination of these early investigators in developing these technics is laudable; many of those used today are based on the early investigations.

There are many ways to study the response of cells to radiation, both in vivo (in the living organism) and in vitro (in glassware). Tissue culture (growing of animal and human cells in a bottle or tube by providing nutrients) is an extremely useful in vitro tool for studying the response of single cell types to radiation. In this situation vasculature and other physiologic factors present in the living organism do not contribute to the response.

Some studies are performed on asynchronous populations of cells; that is, the population being studied contains cells in all four phases of the cell cycle (G_1, S, G_2 and M). Other studies are performed on synchronous populations, necessitating technics that place all or most cells in a given phase of the cell cycle at a given time. Dividing populations of cells in the body are asynchronous; therefore the first method more nearly simulates the in vivo situation. However, the second method permits observations of cellular response in different phases of the cell cycle.

One of the classic studies in radiation biology was the construction of the cell survival curve by Puck and Marcus.[28] These two investigators grew HeLa cells (derived from human carcinoma of the cervix) in tissue culture and kept them alive for many generations. In fact, the first culture was started in 1956 and today it is possible to purchase HeLa cells from biologic supply companies. Puck and Marcus exposed the cells to various doses of radiation and observed the ability of the cells to reproduce. Other investigators (Withers[36] and McCulloch and Till[25]) have developed technics to construct cell survival curves using in vivo systems, such as the skin and hemopoietic systems.

Because the true response of the individual cell cannot be observed in all in vivo systems, this chapter will deal with those responses which have been observed in vitro and in appropriate in vivo systems.

Interphase Death

One response of the cell to radiation is death before the cell enters mitosis. This response is called *interphase death*; synonymous terms are *nonmitotic* or *nondivision death*. Interphase death can occur both in cells that do not divide and are long-lived (e.g., adult nerve) and in rapidly dividing cells (e.g., blood cell precursors in the bone marrow) and has been observed after irradiation of oocytes, erythroblasts and cancer cells.

The dose that produces this response varies with cell type—lymphocytes exhibit interphase death at doses less than 50 rads; mouse spermatogonia, at 25 rads; but 30,000 rads cause interphase death in only 50% of a population of yeast cells. Although the relationship among interphase death, cell type and dose is poorly understood, in general, rapidly dividing, undifferentiated (radiosensitive) cells exhibit interphase death at lower doses than nondividing, differentiated (radioresistant) cells. The one exception to this generalization is the lymphocyte (an RPM cell), which undergoes interphase death at very low doses.

The mechanism that leads to interphase death generally begins within a few hours postirradiation, but death itself does not occur at this time. Within a few days postirradiation, the nucleus exhibits prominent changes, the normal architecture disappears and the nucleus appears as a dark staining area without recognizable chromatin material. The entire cell at this time exhibits changes typical of cell necrosis (cell degradation following death).

The mechanism producing interphase death is obscure; it appears unrelated to mitosis inasmuch as it is not an unsuccessful attempt by the cell to divide. A possible explanation is that interphase death is caused by biochemical changes in the cell, such as a decrease in energy production by mitochondria.

Division Delay

A second response of cells to radiation involves mitosis. In dividing, asynchronous populations of cells, a certain proportion of cells will be in mitosis at any one time. The ratio of the number of cells in mitosis at any one time to the total number of cells in the population is termed the *mitotic index.* If this ratio is plotted on graph paper, with mitotic index plotted on the *y* axis and time plotted on the *x* axis, the mitotic index remains relatively constant—as some cells are completing mitosis, others are entering prophase, maintaining the mitotic index at a status quo.

Irradiation of the population disturbs this constant ratio of mitotic to nonmitotic cells (Fig. 4-1). Cells in mitosis at the time of irradiation complete division, but those about to enter division are delayed in G_2. The mitotic index therefore decreases for a period of time as some

Fig. 4-1.—Effects of radiation on an asynchronous, dividing population of cells. Prior to irradiation, the mitotic index remains at a constant value; radiation *(wavy arrow)* disturbs this constant ratio of cells, effecting a decrease in the number of cells in mitosis *(mitotic delay),* which may be overcome (depending on dose), producing an increase in the number of cells in mitosis at any time *(mitotic overshoot).*

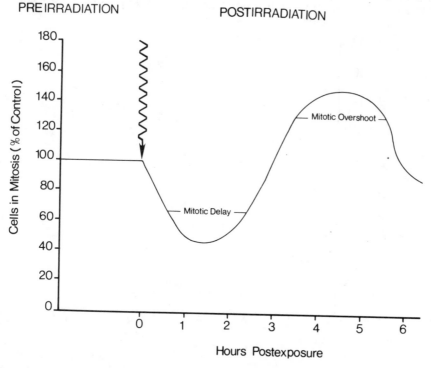

cells are stopped from proceeding through mitosis at their appointed time. If the dose is low enough, these cells recover from this delay and proceed through mitosis, resulting in an increased number of cells in mitosis—appropriately termed *mitotic overshoot*. During this time, cells entering mitosis consist of two classes: those normally progressing through mitosis which were not delayed by irradiation and those which were delayed by irradiation. This cellular response to radiation is termed *division* or *mitotic delay*.

Canti and Spear[13] observed division delay in chick fibroblasts in tissue culture, irradiated with varying doses of γ-rays from a radium source (Fig. 4-2). Low doses (50 rads) produced a negligible effect on mitotic index; as dose increased (83 and 300 rads), the response became more pronounced, i.e., both the length and magnitude of the delay were increased. The mitotic overshoot increased in magnitude with these effects—the length and magnitude of the overshoot reflecting the delay.

The overshoot was followed by a return of the mitotic index to its

Fig. 4-2.—Experimental findings by Canti and Spear illustrating the dose dependence of mitotic delay. See the text for further explanation. (From Canti, R. G., and Spear, F. G.: The effect of gamma irradiation on cell division in tissue culture in vitro, Part II. Proc. R. Soc. Lond. [Biol.] 105:93, 1929. Courtesy of the Royal Society.)

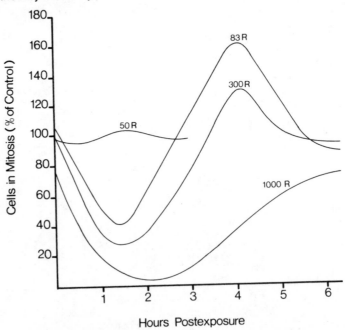

pre-irradiated (100%) value after exposure to low doses (50 and 83 rads). However, after higher doses the mitotic index fell below the pre-irradiated value following the mitotic overshoot and remained there. At these doses a third mechanism of damage was operational— the cells divided but died after division. Termed *reproductive failure*, this response will be discussed in the next section.

A more dramatic response was observed following a dose of 1000 rads. Not only were the length and magnitude of delay greatly increased, but no mitotic overshoot occurred, indicating the cells apparently were not able to overcome the block imposed by the radiation and therefore died before division (interphase death).

Division delay can be induced in cells by doses as low as 10 rads; 50 rads to human kidney cells in tissue culture results in division delay. An opportunity to observe division delay in vivo occurred when five people were overexposed to radiation in a reactor accident (Y-12 accident). Dose estimates to these individuals ranged from 50 to 200 rads. Observation of cells in the bone marrow of these casualties showed a progressive decrease in mitotic index approaching zero on the fourth day postirradiation.

The underlying cause of mitotic delay is unknown; some theories proposed to explain this phenomenon are (1) a chemical involved in division is altered by irradiation, (2) proteins necessary for mitosis are not synthesized or (3) DNA synthesis does not progress at the same rate following irradiation.

In summary, division delay is a dose-dependent phenomenon; the decrease in mitotic index and the length of the delay is a function of dose. At low doses the time of the delay and the decrease in mitotic index is much less than it is at higher doses. The mitotic overshoot reflects the ability of the cells to overcome the radiation-induced block and proceed through mitosis along with unaffected cells, indicating that the process is reversible. Division delay also appears to be a function of cell cycle stage; some stages are more sensitive to radiation than others. Essentially, radiation acts as a synchronizing agent by selectively affecting cells in the most sensitive cell cycle stage and delaying their progression through mitosis. This response will be discussed later in this chapter.

Reproductive Failure

A third type of cellular response to radiation is a decrease in the percentage of cells surviving after irradiation that have retained their reproductive integrity—i.e., are capable of reproducing. This is termed *reproductive failure* and is defined as the inability of the cell to undergo repeated divisions after irradiation. By this definition, all cells that cannot *repeatedly* divide are considered nonsurvivors or "dead" even though they may still be technically alive (i.e., they are metabolizing or capable of a limited number of divisions). This concept can be understood in terms of the target theory (Chapter 2). The ability of a cell to reproduce is directly related to the integrity of the chromosomes. If it is assumed that chromosomes (or DNA) are the critical sites (targets) in the cell and that an ionizing event (hit) occurring *only* in the target is responsible for cell death, then damage to the chromosomes may result in death of the cell. However, much of this damage can be, and usually is, repaired. If it is not repaired, the cell may still retain some ability to divide, doing so one or more times following irradiation.

Puck and Marcus[28] experimentally determined reproductive failure by exposing human cells (HeLa) to various doses of radiation and counting the number of colonies formed by these irradiated cells. They graphically expressed the ability of cells to reproduce after different doses of radiation in the form of a semilogarithmic curve, where dose is plotted on a linear scale (*x* axis) and surviving fraction on a logarithmic scale (*y* axis). This curve, illustrated in Figure 4-3, is termed a *cell survival curve.*

Fig. 4-3.—Illustration of the survival curve determined by Puck and Marcus. Below 150 rads the curve exhibits a shoulder region, becoming exponential at higher doses. (From Puck, T. T., and Marcus, T. I.: Action of x-rays on mammalian cells, J. Exp. Med. 103:653, 1956.)

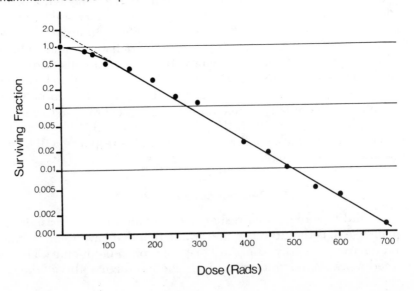

The curve shown is characteristic for the survival of mammalian cells exposed to different doses of x- or γ-rays and can be described by its shape. Below 150 rads, the curve exhibits a "shoulder" region; in this area, equal increases in dose do not cause a corresponding equal decrease in surviving fraction. In other words, doses less than 150 rads are inefficient in producing cell death. This region of the cell survival curve actually indicates that, in mammalian cells, damage must be accumulated in more than one target before death occurs. If there were only one target in each cell, each dose of radiation, starting with the lowest, would be as efficient as the next in causing cell death—the data would be expressed as a straight line on a logarithmic plot from its origination on the y axis, i.e., at 0 dose and 100% surviving fraction (Fig. 4-4).

However, this curve does form a straight line on a logarithmic plot, *but* at doses *greater* than 150 rads. In this region equal increases in dose *do* cause a corresponding equal decrease in surviving fraction, but the absolute number of cells killed varies (Table 4-1). These fractional decreases indicate a logarithmic relationship or exponential function and is the same principle that describes the decay of a radioactive isotope.

The exponential response of cells following irradiation is due to the random probability of radiation interaction in matter, i.e., we are not absolutely certain where an ionization will occur. As a result, if a population of cells with n targets (n being more than one) is irradiated with a given dose, some cells in the population will sustain hits in all the targets (lethal damage), some cells will sustain hits in only a few targets (sublethal damage) and some will not be hit at all. As dose in-

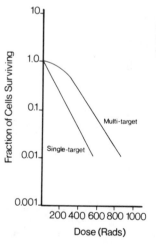

Fig. 4-4.—Survival curve exhibited by cells with a single target and cells with multiple targets.

TABLE 4-1.—TYPICAL DATA RELATING DOSE AND SURVIVING FRACTION*

ORIGINAL CELL NUMBER	DOSE DELIVERED (RADS)	FRACTION (%) CELLS KILLED	NUMBER OF CELLS KILLED
100,000	500	50	50,000
50,000	500	50	25,000
25,000	500	50	12,500
12,500	500	50	6,250
6,250	500	50	3,125

*This table illustrates the exponential response of cells exposed to equal dose increments. Note the same dose (500 rads) always kills the same proportion (50%) of cells, but the absolute number of cells killed varies. This type of data indicates a logarithmic relationship between dose and surviving fraction. Two doses of 500 rads actually inactivate only 75%, not 100%, of the original population. If two doses of 500 rads did inactivate 100% of the population, the response would be linear, not exponential.

creases, there is an increased probability that ionization will occur in previously "hit" areas—because at high doses more hit areas occur than nonhit areas, damage has accumulated in more targets and radiation is now efficient in causing cell death.

The cell survival curve is defined by three graphic parameters: n (extrapolation number); D_0 dose (D_{37} dose); and D_q (quasi-threshold dose) (Fig. 4-5).

The extrapolation number n is determined by extrapolating the linear portion of the curve back to its intersection with the y axis. This term was originally referred to as the target number and was assumed to represent the number of targets that must be hit in each cell to cause cell death. However, objections arose over the use of the term

Fig. 4-5.—Survival curve of cells illustrating the portion of the curve from which n (extrapolation number), D_q (quasi-threshold dose), and D_0 (an expression of radiosensitivity) are derived.

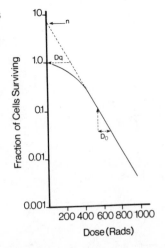

"target," and it is now referred to as the extrapolation number. The n for mammalian cells ranges between two and ten.

D_0 is determined from the exponential portion of the curve and is the reciprocal of the slope (1/slope). This expression was derived mathematically from the target theory and is defined as that dose which inactivates all *but* 37% of the population. D_0 is an expression of radiosensitivity of a population; cells with high D_0 doses are less sensitive (more resistant) than populations with low D_0 doses. D_0 doses for different populations of mammalian cells vary between 100 and 220 rads.

The D_q or quasi-threshold dose defines the width of the shoulder region of the curve and is the dose at which point the survival curve becomes exponential.

The survival curve derived by Puck and Marcus is characteristic for mammalian cells irradiated with x- or γ-rays under standard conditions of pressure and oxygenation. Reproductive failure can be affected by factors such as the LET of the radiation or external conditions under which the cells are irradiated. The effects of these and other factors on cellular response will be discussed in the next section. These cellular responses to radiation are summarized in Table 4-2.

Factors Influencing Response

The response of cells to radiation can be affected by various factors; the effects are either to diminish the response, enhance the response or elicit the response at a different time. Therefore, these factors *appear* to affect the radiosensitivity of cells—cells exhibiting enhanced response appear more radiosensitive; those exhibiting diminished response appear more radioresistant. However, the inherent sensitivity of the cell has not changed; the cell is basically the same with the same characteristics (i.e., dividing or nondividing and differentiated or undifferentiated). What has changed is an external factor such as the LET of the radiation or the environment in which the cell is growing, thus exerting an influence on the response of the cell (or organism) to radiation. But the inherent sensitivity of the cell as determined by Bergonié and Tribondeau remains the same.

These factors are those which Ancel and Vitemberger defined as affecting cellular radiosensitivity, i.e., factors to which the cell is exposed pre- or postirradiation. Because they do not change specific cellular characteristics that determine cell radiosensitivity or resistance, a change in radiation response induced by any of these factors can be termed *conditional sensitivity*.

TABLE 4-2.—TYPES OF CELL DAMAGE

TYPE	CHANGES	DOSE	MECHANISM
Interphase death	Normal nuclear architecture disappears	High, in most cases, except lymphocytes	May be biochemical
Division delay	Lowered mitotic index; cells are delayed from proceeding through mitosis for a given time	Exhibited by dividing cells; dose-dependent, degree of response varying with dose	Unknown: change in a chemical involved in division; proteins are not synthesized; DNA synthesis is affected
Reproductive failure	Loss of reproductive integrity; cell cannot undergo repeated division	Exhibited by dividing cells; dose-dependent, high doses affect greater number	Unknown

The factors affecting response will be discussed in the following sections and are grouped as physical factors, chemical factors and biologic factors (Table 4-3).

Physical Factors

LET and RBE

As discussed in Chapter 2, the quality of the radiation has an effect on the biologic response. Linear energy transfer (LET) is the term that describes the rate at which energy is lost from different types of radiation while traveling through matter. Irradiation of the same biologic system with different LET radiations will produce different qualities of biologic response. The term relating biologic response to the quality of the radiation is *RBE*.

To compare the effect on cell survival of high LET radiations (e.g., neutrons) to low LET radiations (e.g., x-rays), survival curves are constructed of cells exposed to these two types of radiation (Fig. 4-6). RBE can be calculated from this curve by choosing the same survival level on the exponential portion of both curves, e.g., 0.1, 0.01 or the D_0 dose, and comparing the doses from the two types of radiation that produce this response. In the example 600 rads of x-rays and 200 rads of neutrons produce a surviving fraction of 0.1. Therefore, the RBE of neutrons equals 3, which can be interpreted to mean that neutrons are three times more effective than x-rays in killing 90% of a population of cells. In addition, the shoulder region of the curve of cells exposed to high LET radiation is much smaller than that of cells exposed to low LET radiation, indicating damage does not have to be accumulated before cell death occurs. The shoulder region may even be eliminated if the LET is high enough.

The dependence of biologic response on the LET of the radiation can be explained on the basis of energy deposition by different types of radiation; low LET radiation produces sparse ionizations separated by relatively long distances, whereas high LET radiation produces

TABLE 4-3.—FACTORS AFFECTING RESPONSE

PHYSICAL	CHEMICAL	BIOLOGIC
LET	*Sensitizers*	Cell cycle
Dose-rate	Oxygen	Intracellular repair
	Halogenated pyrimidines	
	Others—partial sensitizers	
	Protectants	
	Cysteine	
	Cysteamine	

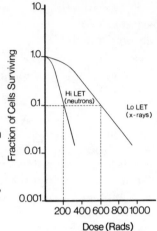

Fig. 4-6.—Comparison of high and low LET radiations on a survival curve. That dose which produces 0.1 fraction of cells surviving with x-rays is 600 rads, but with neutrons it is only 200 rads. In this case, the RBE for neutrons equals 3.

dense ionizations in a very short distance. The ionization density produced by high and low LET radiations in the same population of cells is illustrated in Fig. 4-7. The high LET radiation has produced two hits in the nuclei of two different cells, whereas the low LET radiation has produced only one hit in the nuclei of two cells. If we assume that most mammalian cells contain more than one target that must be hit to cause cell death and that these targets are contained within the nucleus, it can readily be seen that the high LET radiation with its high ionization density is more efficient in producing cell death than the low LET radiation. The decrease or absence of the shoulder region on the survival curve of cells irradiated with high LET radiations is due to the large number of ionizations produced in the targets.

In some systems (e.g., viruses) a single ionization in the target volume will result in death of the organism. In this type of system many of the ionizations from high LET radiations will be wasted; if only one hit is required to kill the organism, two hits will not render it "more dead." Therefore, low LET radiations will be more efficient for producing cell killing than high LET radiations on a dose basis.

Dose-Rate

Another physical factor affecting the response of cells to radiation is dose-rate, i.e., the rate at which the radiation is delivered. A dose-rate effect has been observed in many types of biologic damage, including reproductive failure (Fig. 4-8), division delay, chromosome aberrations (especially complex aberrations) and survival times of organisms exposed to total body irradiation. All studies have shown low dose-rates to be less efficient for producing damage than high dose-rates.

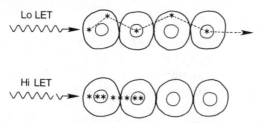

Fig. 4-7.—Comparison of the effects of low and high LET radiations on a population of cells. Note the irregular path of the low LET radiation, interacting with four cells, compared to the relatively straight path of the high LET radiation interacting with only two cells. However, the low LET radiation produces only *one* hit in two nuclei whereas the high LET radiation produces *two* hits in two nuclei.

An explanation for the dependence of biologic response on dose-rate is that low dose-rates allow repair to occur before enough damage has accumulated to cause death of the cell. High dose-rates may not permit repair because of the short time period over which the radiation is given. High LET radiations do not show a dose-rate effect. This is not surprising, due to the dense ionizations produced by high LET radiations that "hit" enough targets and/or the repair mechanisms to kill the cell.

Fig. 4-8.—Survival curves illustrating a dose-rate dependence. The low dose-rate curve exhibits a broader shoulder and more shallow slope when compared to the high dose-rate radiation.

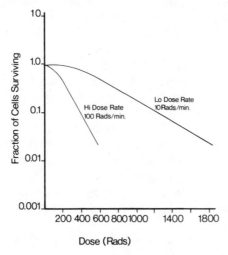

Chemical Factors

Radiation Sensitizers

Many chemicals can change the response of cells to radiation. Some of these chemicals enhance response—these are termed *radiosensitizers*. Those which diminish response are termed *radioprotectors*.

A true radiosensitizer is one that increases the cell-killing effect of a given dose of radiation. Many chemicals have been found that fit this criteria; however, the one that has the most dramatic effect and has been shown to universally (meaning in all organisms tested) enhance radiation response is oxygen. The universality of this radiosensitizer is attested to by the fact that not only does it enhance the radiation response of mammalian cells in tissue culture, but its effects have been observed in all classes of organisms—from bacteria to the whole organism. Because of its universal and dramatic effect, the response of cells to radiation in the presence of oxygen has been given a specific name—the *oxygen effect*.

An interesting facet of the oxygen effect was observed from experiments in which the timing of the delivery of oxygen varied, i.e., oxygen was given either pre-irradiation, postirradiation or simultaneously with radiation. In these situations, oxygen was found to be most effective when administered simultaneously with radiation; oxygen administered either pre- or postirradiation did not produce as dramatic a response. These observations led investigators to postulate the mechanism of the oxygen effect. Because the interaction of radiation with matter is a rapid process and because oxygen was found to be most effective when administered simultaneously with radiation, the effects of oxygen were postulated to involve the reaction of radiation at the chemical level. Although the mechanism of oxygen enhancement of radiation response remains obscure, several theories have been advanced, two of which are most accepted.

The first theory involves the free radicals formed as a result of radiation interaction with the water content of a cell. Believed to be responsible for a large portion of radiation damage, oxygen may either enhance the formation of free radicals or draw the existing free radicals into chain reactions producing new, highly damaging radical species. Another mechanism postulated for the sensitizing properties of oxygen is that many of the chemical changes that occur as a result of irradiation are reversible if oxygen is not present. However, if oxygen is present, it may block these restoration processes, thus increasing damage in the cell.

Figure 4-9 is a comparison of the survival of the same population of cells, one in the presence of oxygen and one in the absence of oxygen. The curve of cells irradiated in oxygen exhibits two changes:

1. The shoulder region of the curve is smaller.
2. The slope of the exponential portion of the curve is steeper, resulting in a decreased D_0 dose.

Because oxygen increases the amount of damage through increased free radical formation or the blocking of restoration processes, more targets are lethally affected per given dose of radiation. This results in an enhanced response reflected by these changes in the survival curve, indicating an increase in radiosensitivity.

The enhancement of the response of cells to radiation in the presence of oxygen does not increase in an unlimited fashion (Fig. 4-10). Oxygen concentration is measured by the pressure it exerts and is termed *oxygen tension*; the units of measurement are millimeters of mercury. The response occurs with oxygen tensions between 0 and 20 mm Hg; those greater than 20–40 mm Hg (the oxygen tension in air at sea level) do not result in a further increase in radiation sensitivity.

The term that compares the response of cells or organisms to radiation in the presence and in the absence of oxygen is the *oxygen enhancement ratio (OER)*. OER is defined as the dose of radiation that produces a given biologic response in the absence of oxygen divided by the dose of radiation that produces the same biologic response in

Fig. 4-9.—Survival curve of two identical cell populations exposed to various doses of x-rays under aerated and hypoxic conditions. The D_0 for hypoxic cells is 300 rads, whereas the D_0 for oxygenated cells is 100 rads; therefore, the OER equals 3.

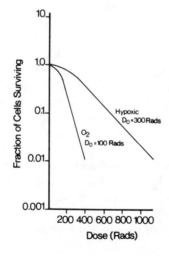

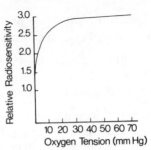

Fig. 4-10.—Curve illustrating the limitations of the oxygen effect with increasing oxygen tension. Note the plateau in sensitivity at oxygen tensions between 20 and 40 mm mercury (the oxygen tension in air at sea level).

the presence of oxygen. For example, if 300 rads is the D_0 of a population of cells exposed under hypoxic conditions (decreased oxygen tension) and 100 rads is the D_0 of the population when oxygen tension is increased, the OER is 300/100, or 3. For mammalian cells, the OER is between 2 and 3 and is generally given as approximately 2.5.

The oxygen effect is most pronounced with x- and γ-rays (low LET radiations) and is not as effective with neutrons and alpha particles (high LET radiations) (Fig. 4-11).

Readily explainable by the physical differences between the two types of radiations, the amount of damage produced by high LET radiations would not be reparable, therefore the presence of oxygen would not enhance the radiation response to the same extent as with low LET radiations. The OER for high LET radiations varies between 1.7 and 1.2.

The role of oxygen in the treatment of tumors with radiation has become very important over the past 15 years. This will be discussed further in Chapter 10.

Other Sensitizers

In addition to oxygen, other compounds also are known to sensitize cells to radiation. Some of these are halogenated pyrimidines, actinomycin D, hydroxyurea and vitamin K.

Halogenated pyrimidines are chemical compounds that substitute for the base thymidine in the DNA molecule. Two halogenated pyrimidines that are effective radiosensitizers are 5-bromodeoxyuridine (5-BUDR) and 5-iododeoxyuridine (5-IUDR). These two compounds, if present in a cell, will be selectively incorporated into DNA in place of thymidine changing the molecule and thereby rendering it more susceptible to radiation damage. Because the radiation-enhancing effects of these two compounds is dependent on their incorporation into DNA, unlike oxygen, they must be present for several cell cycles before irradiation.

Both these compounds sensitize cells by a factor of 2; in other

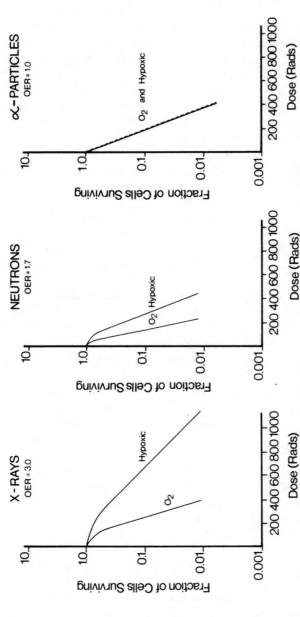

Fig. 4-11.—Comparison of the oxygen effect with x-rays, neutrons and α-particles. Note it is most pronounced with x-rays and least pronounced with α-particles.

words, if BUDR is incorporated into DNA, it will take one-half the dose to produce the same response as is produced in cells without BUDR. The presence of BUDR has been shown to enhance two cellular responses to radiation: reproductive failure and division delay.

Many other chemical compounds have been examined to determine their ability to sensitize cells to radiation. Some found to be partial radiosensitizers are the antibiotic actinomycin D, hydroxyurea, synkavit (vitamin K), puromycin and 5-fluorouracil, all of which have been used in combination with radiation for the treatment of cancer. However, the only radiosensitizing agent that has gained widespread use in clinical radiation therapy is oxygen.

Radiation Protectors

Approximately 20 years ago it was discovered that certain compounds, if present at the time of radiation, had a protective effect on the organism. It was also observed that these compounds had to be present at the time of irradiation to exert the protective effect—if administered immediately following irradiation, no protective effect was noted. These compounds, radioprotectors, act by reducing the effective dose of radiation to the cells; for this reason they are often termed *dose modifying compounds.*

One group of compounds that have radioprotectant properties are chemicals that contain a sulfhydryl group (sulfur and hydrogen bound together, designated SH). Two amino acids in the body belonging to this group of sulfhydryl compounds are cysteine and cysteamine. In fact, cysteine was one of the first compounds found to have radioprotectant properties.

When one of these compounds is given prior to radiation, a larger dose is necessary to produce the same response as when the compound is not present. The term that relates this difference in response to the presence of the protective compound is *dose reduction factor (DRF).*

DRF is defined as the ratio of the radiation dose necessary to produce a given effect in the presence of a protecting compound to the radiation dose necessary to produce the same effect in the absence of the same compound. The DRF for the sulfhydryl-containing compounds is approximately 1.5 to 2.0, i.e., if the compound is present during radiation, almost twice the dose is required to produce the same response as that produced by one-half the dose in the absence of the compound (Fig. 4-12).

Many hypotheses have been advanced concerning the mechanism of action of these dose-modifying agents. The most generally accepted hypothesis today is that these agents protect either by competing for the radiation-produced free radicals or by giving up a hy-

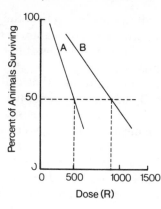

Fig. 4-12.—Effects of the administration of a radioprotectant compound on the survival of irradiated animals. In the presence of the compound (B) 1000 R of total body irradiation is necessary to reduce the percent survivors 50%, whereas in the absence of the compound (A), 500 R produces this response. Therefore, the dose reduction factor (DRF) in this example equals 2.

drogen atom to ionized molecules in the cell, neutralizing the effects of radiation and restoring the molecule to its original pre-irradiated state.

The sulfhydryl compounds are most efficient with x- and γ-rays and have a negligible effect with high LET radiations such as alpha particles and neutrons. Thus the protection observed with these compounds parallels the oxygen effect; these compounds exhibit maximum protection with low LET radiation and minimum protection with high LET radiations, just as oxygen exhibits maximum sensitivity with low LET and minimum sensitivity with high LET radiations.

Other compounds that offer protection do so by producing systemic hypoxia, i.e., they reduce the amount of oxygen available to a cell. These compounds have not been put into widespread use because their effectiveness remains questionable.

The modifying agents discussed above are not in widespread use due to a number of factors. It would seem feasible that radioprotectant compounds would be useful in radiation safety for protection against overexposure. However, the concentration of these compounds necessary to protect an organism is toxic, therefore prohibiting their use. In addition, these agents must be present in the cell at the time of irradiation, not postexposure, making them useless for treatment in cases of accidental overexposure.

Biologic Factors

Cell Cycle

In addition to chemical and physical factors that alter the cellular radiation response, there are some important biologic factors that also modify the response. One biologic factor that has a great influence on cellular response is the position of the cell in the cell cycle at the time of irradiation. By using technics that synchronize a population of cells, i.e., place all cells in one phase of the cell cycle, it is possible to observe the response of cells in different phases.

A summary of the experimental findings indicates that cells are more radiosensitive when irradiated in G_2 and M, less sensitive in G_1 and least sensitive (most radioresistant) during DNA synthesis (Fig. 4-13). In general M is considered to be the most radiosensitive phase in the cell cycle and S the most resistant.

Division delay, a dose-dependent cellular response, is directly related to the position of the cell in the cell cycle. Low doses of radiation affect cells in the most radiosensitive phases of the cell cycle (G_2 and M), delaying their progress through mitosis for a given period of time. Higher doses, affecting cells in all phases of the cycle, both radiosensitive and radioresistant, produce a longer mitotic delay.

Intracellular Repair

A second biologic factor that influences response is the capability of cells to repair sublethal damage, i.e., recover from radiation injury. Cell survival curves exhibit an initial shoulder region, indicating that damage must be accumulated in more than one target before cell death occurs.

In a series of experiments with cells in tissue culture, Elkind and

Fig. 4-13.—Graph illustrating the effect of cell cycle position on cell survival of a synchronous population of cells. M and G_2 are the most radiosensitive, whereas early S (ES) and late S (LS) are the most resistant. (From Sinclair, W. K.: Cyclic responses in mammalian cells in vitro, Radiat. Res. 33:620, 1968.)

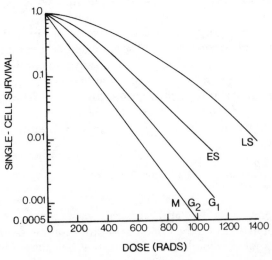

co-workers[14] determined that when the same total dose is administered in fractions separated by a period of time, the number of cells
surviving increased with the time between fractions. In addition, the
survival curve after the second dose exhibited the same D_0, n, and D_q
as the survival curve following the first dose of radiation (Fig. 4-14).

In essence, cells surviving the first dose fraction respond as unirradiated cells to the second fraction. This observation was interpreted
as meaning that radiation damage had been repaired between the first
and second doses, indicating that cells have the capability of re-
covering from sublethal injury (injury not resulting in death). For this
reason a higher total dose is necessary to produce the same biologic
response when the dose is fractionated than when given acutely. In
addition, this process appears to be completed in the cell within 24
hours postirradiation.

Of particular interest is the influence of oxygen on repair. Experimentally, a hypoxic environment appears to hinder the cell's ability to
repair sublethal damage, specifically when compared to well-oxygenated cells.

The ability of the cell to repair intracellular radiation damage has
implications for radiation therapy, both in terms of the time required
for repair and of the effects of hypoxia on this phenomenon. These
topics will be discussed further in Chapter 10.

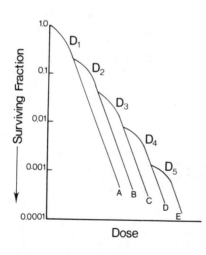

Fig. 4-14.—The effect of dose
fractionation on cell survival. After
each succeeding dose fraction
(D_1, D_2, D_3, etc.) the curve exhibits
the same shoulder, slope and
extrapolation number indicating
that, within a given period of time,
damage is repaired. (From Elkind,
M. M., and Whitmore, G. F.:
Radiobiology of Cultured
Mammalian Cells [New York:
Gordon & Breach, 1967].)

REFERENCES

1. Arena, Victor: *Ionizing Radiations and Life* (St. Louis: C. V. Mosby, 1971).
2. Bacq, Z. M., and Alexander, P.: *Fundamentals of Radiobiology* (2nd ed.; New York: Pergamon Press, 1961).
3. Barendsen, G. W., *et al.*: Effects of different ionizing radiations on human cells in tissue culture. II. Biological experiments, Radiat. Res. 13:841, 1960.
4. Barendsen, G. W.: Impairment of the proliferative capacity of human cells in culture by alpha particles with differing linear energy transfer, Int. J. Radiat. Biol. 8:453, 1964.
5. Bedford, J. S., and Hall, E. J.: Survival of HeLa cells cultured in vitro and exposed to protracted gamma irradiation, Int. J. Radiat. Biol. 7:377, 1963.
6. Belli, J. A., *et al.*: Radiation recovery response in mammalian tumor cells in vivo, Nature 211:662, 1966.
7. Belli, J. A., *et al.*: Radiation response of mammalian tumor cells. I. Repair of sublethal damage in vivo, J. Natl. Cancer Inst. 38:673, 1967.
8. Berry, R. J., and Cohen, A. B.: Some observations on the reproductive capacity of mammalian tumor cells exposed in vivo to gamma radiation at low dose rates, Br. J. Radiol. 35:489, 1962.
9. Berry, R. J.: Modification of radiation effects, Radiol. Clin. North Am. 3:249, 1965.
10. Berry, R. J., *et al.*: Reproductive capacity of mammalian tumor cells irradiated in vivo with cyclotron produced fast neutrons, Br. J. Radiol. 38:613, 1965.
11. Bleehen, N. M.: Combination therapy of drugs and radiation, Br. Med. Bull. 29:54, 1973.
12. Brown, J. M., *et al.*: Preferential radiosensitization of mouse sarcoma relative to normal skin by chronic intra-arterial infusion of halogenated pyrimidine analogs, J. Natl. Cancer Inst. 47:75, 1971.
13. Canti, R. G., and Spear, F. G.: The effect of gamma irradiation on cell division in tissue culture in vitro, Part II, Proc. R. Soc. Lond. [Biol.] 105:93, 1929.
14. Elkind, M. M., and Sutton-Gilbert, H.: Radiation response of mammalian cells grown in culture. I. Repair of x-ray damage in surviving Chinese hamster cells, Radiat. Res. 13:556, 1960.
15. Elkind, M. M., *et al.*: Actinomycin D; suppression of recovery in x-irradiated mammalian cells, Science 143:1454, 1964.
16. Elkind, M. M., *et al.*: Radiation response in mammalian cells in culture. V. Temperature dependence of the repair of x-ray damage in surviving cells (aerobic and hypoxic), Radiat. Res. 25:359, 1965.
17. Field, S. B.: The relative biological effectiveness of fast neutrons for mammalian tissues, Radiology 93:915, 1969.
18. Hall, E. J., and Bedford, J. S.: Dose-rate: Its effect on the survival of HeLa cells irradiated with gamma rays, Radiat. Res. 22:305, 1964.
19. Hall, E. J.: Radiation dose rate: A factor of importance in radiobiology and radiotherapy, Br. J. Radiol. 45:81, 1972.
20. Hornsey, S.: *Advances in Radiobiology* (New York: Academic Press, 1957).
21. Howard-Flanders, P., and Alper, T.: The sensitivity of micro-organisms to radiation under controlled gas conditions, Radiat. Res. 7:518, 1957.

22. Lea, D. E.: *Actions of Radiations on Living Cells* (2nd ed., Cambridge: Cambridge University Press, 1962).
23. Littbrand, B., and Revesz, L.: Effects of a second exposure to x-rays on post-irradiation recovery, Nature 214:841, 1967.
24. Littbrand, B., and Revesz, L.: The effect of oxygen on cellular survival and recovery after radiation, Br. J. Radiol. 42:914, 1969.
25. McCulloch, E. A., and Till, J. E.: The sensitivity of cells from normal mouse bone marrow to gamma radiation in vitro and in vivo, Radiat. Res. 16:822, 1962.
26. Mohler, W. C., and Elkind, M. M.: Radiation response in mammalian cells grown in culture. III. Modification of x-ray survival of Chinese hamster cells by 5-bromo-deoxyuridine, Exp. Cell Res. 30:481, 1963.
27. Nakken, K. F.: Radical Scavengers in Radioprotection, in Ebert, M., and Howard, A. (eds.): *Current Topics in Radiation Research*, (Amsterdam: North Holland Publishing Company, 1965).
28. Puck, T. T., and Marcus, T. I.: Action of x-rays on mammalian cells, J. Exp. Med. 103:653, 1956.
29. Read, J.: Mode of action of x-ray doses given with different oxygen concentrations, Br. J. Radiol. 25:335, 1952.
30. Read, J.: The effect of ionizing radiation on the broad bean root. XI. The dependence of alpha ray sensitivity on dissolved oxygen, Br. J. Radiol. 25:651, 1952.
31. Sinclair, W. K., and Morton, R. A.: X-ray sensitivity during the cell generation cycle of cultured Chinese hamster cells, Radiat. Res. 29:450, 1966.
32. Sinclair, W. K.: Radiation Survival in Synchronous and Asynchronous Chinese Hamster Cells in Vitro, in *Biological Aspects of Radiation Quality*, Proceedings of the Second IAEA Panel, Vienna, 1967 (Vienna: IAEA, 1968).
33. Sinclair, W. K.: Cyclic responses in mammalian cells in vitro, Radiat. Res. 33:620, 1968.
34. Sinclair, W. K.: Dependence of Radiosensitivity Upon Cell Age, in *Time and Dose Relationships in Radiation Biology as Applied to Radiotherapy*, Proceedings of the Conference, Carmel, California, September 1969, BNL Report 50203-C-57. Biology in Medicine TID 4500 (Upton, New York: Brookhaven National Laboratory, 1969).
35. Terasima, R., and Tolmach, L. J.: X-ray sensitivity and DNA synthesis in synchronous populations of HeLa cells, Science 140:490, 1963.
36. Withers, H. R.: Recovery and repopulation in vivo by mouse skin epithelial cells during fractionated irradiation, Radiat. Res. 32:227, 1967.
37. Wright, E. A., and Howard-Flanders, P.: The influence of oxygen on the radiosensitivity of mammalian tissues, Acta Radiol. (Stockholm) 48:26, 1957.

5/Systemic Radiation Response

Systemic radiation response (injury) includes the response of various systems and the total body to radiation. The response of a system is dependent on the sensitivity of the individual organs that comprise the system. In turn, the response of an organ is dependent on the sensitivity of its component tissues, both parenchyma and stroma, and the cell populations comprising these tissues. The basic concepts of radiation sensitivity and cellular response discussed in previous chapters can now be applied to organ response to radiation.

Because preceding classifications of organ sensitivity have been relative (i.e., a comparison of the same radiation-induced biologic effect in the same period of time in two different organs), no unit or value of dose was assigned. In this chapter dose values will be assigned to response, because the severity, type of response and time of appearance is dependent on this factor. To avoid confusion, doses will be defined as low (0 to 100 rads), moderate (100 to 1000 rads) and high (greater than 1000 rads) and will be assumed to be given as a single dose, unless otherwise stated, to the specific organ only. The response of each system and organ will be related to the three medical specialties that utilize radiation (diagnostic radiology, nuclear medicine and radiation therapy); further discussion of each of these clinical areas will be found in Chapters 8, 9 and 10, respectively.

The response of the total body to an acute dose of radiation will be discussed in the next chapter.

What Is "Response"?

The response of a system or organ to radiation can be defined as visible or detectable morphologic and/or functional changes produced by a given dose within a given period of time, indicating that response is a function of both dose and the postirradiation time of evaluation. The majority of this chapter will focus on morphologic response and only the *related* functional changes; functional response per se will be discussed briefly in a separate section.

In most cases the visible effects of radiation on the morphology of an organ are not unique—without the knowledge that radiation exposure has occurred, the observed changes would not implicate radiation as the causative agent. Many other types of trauma will produce the same changes.

The morphologic response of an organ to radiation occurs in two general phases—early and late morphologic changes. Early changes, occurring within six months postirradiation, are a result of cellular damage such as division delay, reproductive failure and interphase death in both the parenchyma and stroma. These changes may be reversible or irreversible, dependent on dose.

Late morphologic changes (occurring later than six months postexposure) are a consequence of irreversible and subsequently progressive early changes. These changes, ranging from minimal to severe, are themselves permanent, irreversible and usually progressive. In general, late response is a function of the type of healing that occurs in the organ.

Healing

Cellular recovery from radiation injury results in healing of radiation damage in an organ. Organ healing occurs by one of two means: *regeneration*, the replacement of damaged cells in the organ by the same cell type present before radiation, or *repair*, the replacement of the original cells by a different cell type. In this latter situation, repair is accomplished by the formation of a scar and is called fibrosis. Regeneration results in a total or partial reversal of early radiation changes, actually restoring the organ to its pre-irradiated state both morphologically and functionally. Therefore, few if any residual late changes are exhibited by the organ.

On the other hand, irreversible early radiation changes heal by repair. This process does not restore the organ to its pre-irradiated condition, thus producing a late radiation response. Repair of radiation damage usually does not contribute to the ability of the organ to perform its function.

Healing of either type is not an absolutely certain event and, under conditions that produce massive and extensive damage, may not occur, resulting in tissue death, i.e., necrosis.

The type of healing and therefore response that occurs in an organ following radiation is a function of both dose and the specific organ irradiated. Although repair can occur in any organ whether radiosensitive or radioresistant, regeneration occurs after low, moderate and even high doses in those organs whose cells are either actively dividing or retain the capability of division, such as skin, small intestine and bone marrow (radiosensitive organs). In these organs repair occurs *only* after those high doses that destroy large numbers of parenchymal cells rendering regeneration impossible. On the other hand, radioresistant organs consisting of nondividing cells (e.g., muscle and CNS) have minimal regenerative capabilities, therefore moderate and high doses result in predominant repair. Low doses to resistant organs produce minimal morphologic damage and therefore minimal response.

A third determinant factor in evaluating the response of an organ to radiation is time. Although it is not difficult to understand why response is dose dependent, the variation of response with time of evaluation needs further clarification. Relatively sensitive organs will respond (show changes) sooner than resistant organs exposed to the same dose; in fact, sensitive organs may manifest a severe response while a minimal response may be observed in resistant organs exposed to the same dose and observed at the same time postexposure. However, at a later time the reverse situation may be true. For example, irradiation of one lung (relatively resistant) and the overlying skin (relatively sensitive) with a single dose of 2000 rads produces marked changes (severe response) in the skin at six months postexposure, but lung changes are less severe. Observations at one year postexposure reveal minimal skin changes, but now the irradiated lung exhibits severe changes, i.e., the lung now appears more sensitive than the skin. Partially due to the inherently slow progression of vascular induced changes in resistant organs, this apparent contradiction in sensitivity and response is directly related to the time of evaluation and the functioning of regenerative processes in the skin and repair processes in the lung. For these reasons resistant organs may ultimately exhibit a more severe response when exposed to the same or even lower doses than radiosensitive organs. Response then is a function of healing which, in turn, is dose-, time- and organ-dependent with time an important factor in evaluation.

Clinical Factors Influencing Response

Because the response of each organ will be related to the three medical specialties using radiation, a word is necessary at this point concerning the doses received from each of these specialties. Low doses (less than 10 rads) are generally delivered to only a portion of the patient's body by diagnostic radiography, fluoroscopy and nuclear medicine. Patient doses in radiation therapy are much higher, usually on the order of 4000 to 6000 rads. However, the total dose in radiotherapy is fractionated; it is split into many small daily doses administered over a period of time (generally 200 rads/day over four to six weeks—a "standard" fractionation schedule). Although the basis of dose fractionation in radiotherapy will be thoroughly discussed in

Chapter 10, it is sufficient for our purposes here to realize that a dose administered in multiple fractions is biologically less effective than a single dose of the same magnitude (Chapter 4). In other words, most organs show less response if the total dose is fractionated than if given as a single dose. This is an extremely important point to keep in mind throughout this chapter to avoid misunderstandings about organ response in radiotherapy.

Another vital factor in understanding clinical organ response is the relationship of volume. In general, although irradiation of part of an organ will elicit the same morphologic response in that specific area as in the whole organ exposed to the same dose, these two situations will have different consequences to the life of the individual. Irradiation of the whole organ may be life-threatening, i.e., the resultant damage may be inconsistent with life. However, a sufficient amount of undamaged organ may remain after only partial irradiation to ensure function and therefore life of the individual. Since this factor is of obvious clinical concern, particularly in radiotherapy, volume effects will be briefly discussed when appropriate.

Response applies not only to the patient but to all individuals working in these three clinical areas. Response will be discussed in relation to persons occupationally exposed to chronic low doses (over a long period of time).

General Organ Changes

In most cases both radiosensitive and radioresistant organs exhibit early and late morphologic changes and are capable of both regeneration and repair. Although the response of these two general classes of organs varies with time and dose, in most organs early changes generally consist of inflammation, edema, and hemorrhage. Higher doses are required to produce these changes at this time in radioresistant than in radiosensitive organs. When irreversible, early changes progress to late changes such as fibrosis, atrophy (decrease in size of an organ) and ulceration. Necrosis, a very severe late response, is the result of failure to repair damage by any means (Fig. 5-1).

Specific organ response will be discussed by system; because the general response of a system is determined by the most radiosensitive organ in that system, attention will be focused on that organ, or organs that account for the changes.

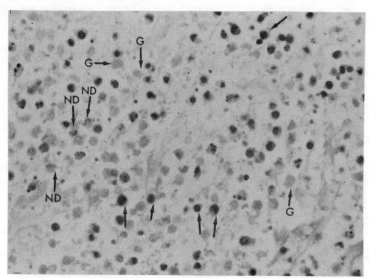

Fig. 5-1.—Photomicrograph of necrosis in an animal tumor (Walker 256) exhibiting hyperchromatic, pyknotic nuclei *(arrows),* the probable remnants of dead lymphocytes and plasma cells. Note "nuclear dust" *(ND)* and "ghost" *(G)* cells. (H & E stain, magnification × 300.)

Hemopoietic System

The hemopoietic system includes the bone marrow, circulating blood, lymph nodes, spleen and thymus (these last three organs are termed *lymphoid* organs).

Bone Marrow

Bone marrow tissues include the parenchymal cells of the marrow, consisting of precursor (stem) cells to cells in the circulating blood and fat cells, and a connective tissue stroma. There are two types of marrow in the adult—red and yellow marrow. Red marrow contains a large number of stem cells in addition to fat cells (Fig. 5-2) and is primarily responsible for supplying mature, functional cells to the circulating blood. In adults, red marrow is present in the following sites: ribs, ends of long bones, vertebrae, sternum and skull bones. Yellow marrow, consisting primarily of fat cells with very few stem cells, is not active in supplying mature cells to the circulating blood and, due to the fat content, is commonly termed "fatty" marrow.

As opposed to the adult where red marrow is located in specific sites, fetal bone marrow is predominantly red marrow.

The primary effect of radiation on the bone marrow is to decrease the numbers of stem cells. Low doses result in a slight decrease with recovery (i.e., stem cell repopulation of the marrow) occurring within

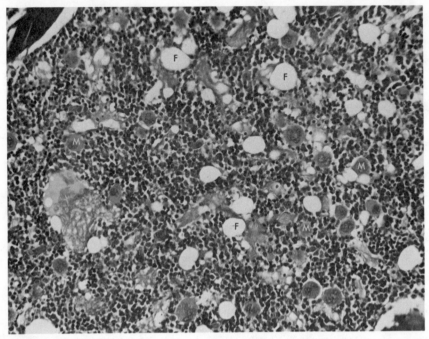

Fig. 5-2.—Photomicrograph of normal bone marrow from rat sternum showing both fat cells *(F)* and megakaryocytes *(M)*; a special stain, Giemsa, would be necessary to identify the remaining red and white cells present. (H & E stain, magnification × 250.)

a few weeks postexposure. Moderate and high doses produce a more severe depletion of cells in the bone marrow, resulting in either a longer period of recovery (time before repopulation of the marrow is complete) and/or less recovery—manifested as a permanent decrease in stem cell numbers and an increase in the amount of fat and connective tissue (Fig. 5-3).

Although all stem cells in the bone marrow are very radiosensitive, variations in sensitivity exist among these different cells. Erythroblasts (precursor cells for red blood cells) are the most radiosensitive, myelocytes (precursors for some white blood cells) are second in sensitivity and megakaryocytes (precursors for platelets) are the least radiosensitive. This variation in sensitivity is manifested as a difference in time and depression of counts in the different stem cells as follows: erythroblasts decrease first and return to normal approximately one week after a moderate dose, myelocytes are depressed in the same time period as erythroblasts but require a longer time to recover (two to six weeks) and depression of megakaryocytes occurs at one to two weeks postexposure requiring a recovery time of two to six weeks.

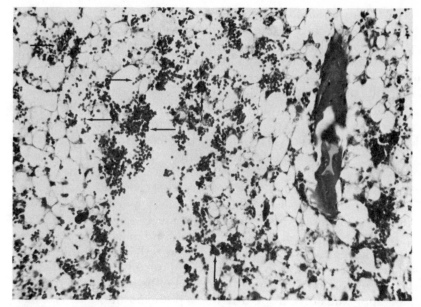

Fig. 5-3.—Photomicrograph of bone marrow from rat sternum exposed to total body dose of 1000 rads, 5 days postirradiation, exhibiting overall hypocellularity and relative increase in fat cells. Note the absence of megakaryocytes, clumps of hyperchromatic, pyknotic cells *(arrows)* and red blood cells *(arrows),* a result of hemorrhage. (H & E stain, magnification × 250.)

Low doses result in decreased stem cell numbers and fast recovery, whereas a more severe decrease in numbers in all cell lines occurs after moderate and high doses with either slow recovery or incomplete recovery of cell numbers relative to pre-irradiated values.

Circulating Blood

With the exception of lymphocytes, the cells in the circulating blood are resistant to radiation (they are nondividing, differentiated cells). However, the circulating blood reflects radiation damage in the bone marrow; as the number of stem cells in the marrow decreases, a corresponding decrease will be exhibited in the number of the respective, mature circulating cells.

The reflection of bone marrow damage in circulating blood cells is dependent on two factors:
1. The sensitivity of the different stem cells.
2. The lifespan of each cell line in the circulating blood.
Although both these factors are important, the latter is more significant in terms of the time of appearance of changes in the circulating blood. All cells in the circulating blood have a finite lifespan (i.e., at certain times they die and must be replaced), varying on the average

from 24 hours (granulocytes) to 120 days (erythrocytes). Damage to the respective stem cells in the bone marrow will be reflected in the circulating blood only when the mature cells die and must be replaced.

The order of depression of blood cell counts as a function of dose and time is lymphocytes decrease first (counts are affected by doses as low as 10 rads), neutrophils are second (doses of 50 rads are necessary to produce a decrease) and platelets and RBC's are third (doses greater than 50 rads; Fig. 5-4). Doses in the low range produce a slight depression of lymphocytes, followed by recovery and a return of the lymphocyte count to pre-irradiated values. Lymphocyte counts will approach zero within a few days following a moderate dose with full recovery occurring within a few months postexposure.

Neutrophil counts fall to minimal values approximately one week following a moderate dose. However, recovery begins soon and neutrophil counts approach normal values within a month postexposure.

The lower doses in the moderate range will have a minimal effect on platelets and RBC's, but the higher doses of this range result in marked depression of these cells. Recovery begins later in these cell lines, approximately the fourth week postexposure, and is usually complete within a few months.

A decrease in the numbers of these various cells has implications for life. Neutrophils and lymphocytes are part of the body's defense mechanism and are important in fighting infection; a decreased number of these cells increases the individual's susceptibility to infection. A decrease in platelets (necessary for blood clotting) results

Fig. 5-4.—Illustration of decrease in number of various blood cells in circulating blood of rat exposed to a moderate dose of total body irradiation. Note the order and time of depletion and recovery of the four cell lines. (From Casarett, A. P.: *Radiation Biology* [Englewood Cliffs, N.J.: Prentice-Hall, 1968].)

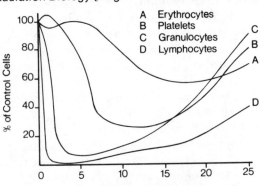

Time in Days after Radiation

in hemorrhage. Anemia follows depression of RBC's and is compounded by hemorrhage throughout the body.

DIAGNOSTIC RADIOLOGY.—Doses in the diagnostic range pose no *major* hazard to the blood and blood-forming organs of the patient or occupationally exposed personnel in terms of decreased cell counts. However, chromosome changes have been observed in circulating lymphocytes following doses in this range.

NUCLEAR MEDICINE.—Because radionuclides are primarily given intravenously for diagnostic purposes, the circulating blood receives an exposure to radiation. Chromosomal changes may occur as a result of these doses, but the probabilities are small because of the small amount of radionuclide given and the consequent low dose to which the blood is exposed. In addition, any chromosome changes in the circulating cells will not be propagated due to the finite lifespan of the cells involved. Whether changes occur in stem cells is not known.

RADIATION THERAPY.—Doses in the therapeutic range will cause a depression of all blood cells, particularly white cells. For this reason blood counts should be routinely obtained on all patients in radiotherapy, particularly those receiving radiation to large volumes of tissue, e.g., in the treatment of Hodgkin's disease and ovarian cancer, or those in which the treatment field includes a large amount of red marrow.

Skin

The skin consists of an outer layer (epidermis), a layer of connective tissue (dermis) and a subcutaneous layer of fat and connective tissue. The skin is supplied with nutrients by blood vessels and contains specialized structures, e.g., hair follicles, sebaceous glands and sweat glands, which arise in the dermis.

The epidermis is made up of layers of cells consisting of both mature, nondividing cells (at the surface) and immature, dividing cells (at the base of the epidermis—the "basal layer"). Cells are periodically lost from the surface of the skin and must be replaced by division of cells in the basal layer. The specific characteristics of the basal cells in the epidermis render the skin sensitive to radiation.

Early observable changes in the skin following a moderate or high dose of radiation are inflammation, erythema (redness of the skin) and dry or moist desquamation (denudation of the skin surface). The skin erythema produced by radiation is not unlike that seen after prolonged exposure to the sun. Produced by an acute dose of 1000 rads, this particular reaction was at one time used as a yardstick to measure the amount of radiation to which an individual had been exposed. The term denoting that dose of radiation which causes this skin erythema is the *skin erythema dose (SED)*.

Moderate doses permit healing to occur in the epidermis by regenerative means resulting in minimal late changes. However, late changes such as atrophy (thinning of the epidermis, Fig. 5-5), fibrosis, decreased or increased pigmentation, ulceration, necrosis and cancer (the latter appearing many years postexposure) may be seen after exposure to high doses (Fig. 5-6).

Accessory Structures

Hair follicles, as an actively growing tissue, are radiosensitive with moderate doses causing a temporary epilation or alopecia (synonyms for hair loss), while high doses may cause permanent epilation.

Sebaceous and sweat glands are relatively radioresistant; damage after high doses produces glandular atrophy and fibrosis resulting in minimal or no function (Fig. 5-5).

DIAGNOSTIC RADIOLOGY.—Doses from diagnostic radiographic and fluoroscopic procedures today pose no hazard in terms of the above described changes (providing, of course, that proper precautions are taken). However, many of these changes, particularly erythema and cancer, occurred on the hands of pioneer workers in radiology. There are known cases of erythema produced in patients as a result of failure to place filtration in the beam.

NUCLEAR MEDICINE.—No early or late skin changes have been observed on the hands of occupationally exposed persons in nuclear medicine, but hand doses are increasing due to both an increase in the number of procedures performed and the amount of certain radionuclides used, e.g., ^{99m}Tc.

RADIATION THERAPY.—Both the early and late changes described above have been observed in patients receiving radiation therapy, particularly when orthovoltage units were in widespread use. With today's high energy units, late skin changes are minimal; fractionated doses in the range of 6000 rads in six weeks usually produce only atrophy of the irradiated area. Although atrophy does decrease the ability of the irradiated area to withstand trauma of any type, this is a relatively minor effect which can be circumvented if further treatment of the area is necessary. Severe necrosis is extremely rare today and is never produced deliberately in practice. It may occur in an individual who is particularly sensitive to radiation, due to unusual treatment protocols or to failure to follow the prescribed treatment plan, e.g., wedges not placed in the radiation beam when specified.

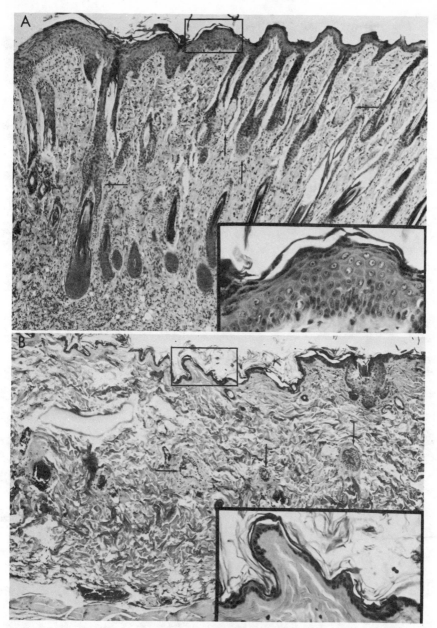

Fig. 5-5.—Photomicrograph of normal **(A)** and irradiated **(B)** rat skin exposed to 2000 rads to a localized area of the body; the animal was sacrificed one year postexposure. Inserts show magnified views of the epidermis. Note the number of cell layers in the epidermis of the unirradiated animal, while that of the irradiated animal consists of only one cell layer. Normal skin contains numerous hair follicles and sebaceous glands *(arrows),* but the irradiated area contains only fibrotic accessory structures *(arrows).* (H & E stain; magnification × 40; inserts, magnification × 500.)

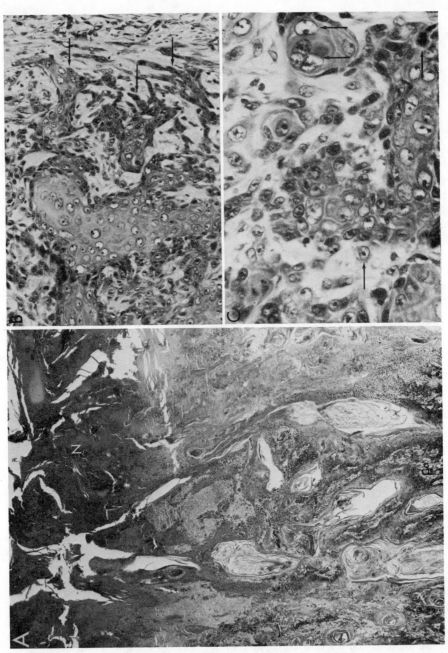

Fig. 5-6.—A, ulcerated, necrotic *(N)* rat skin which had been exposed to 2000 rads to a localized body area and sacrificed one year postexposure. Beneath the ulcerated area is a carcinoma *(Ca)* showing **(B)** invasion *(arrows)* and **(C)** cells with atypical nuclei and mitotic figures *(arrows)*. (H & E stain; A, magnification × 40; B, magnification × 150; C, magnification × 350.)

Digestive System

The alimentary canal consists, in part of the mouth, esophagus, stomach, small intestine, large intestine and rectum; the system is a closed tube throughout the body lined by a mucous membrane. The mucous membrane lining the organs of the digestive tract (like the skin) contains layers of cells, some of which are dividing and undifferentiated (radiosensitive) and some of which are nondividing and differentiated (radioresistant).

Moderate to high doses of radiation produce inflammation in the mucous membranes of the oral cavity (mucositis) and esophagus (esophagitis); however, healing occurs with minimal late changes after moderate doses while high doses result in atrophy, ulceration, fibrosis and esophageal stricture.

The stomach appears to be more sensitive than the esophagus with moderately high doses producing ulceration, atrophy and fibrosis.

The small intestine is the most radiosensitive portion of the GI tract. The lining of the small intestine forms finger-like projections (villi) which aid in the absorption of digested materials into the bloodstream. The cells of this lining are nondividing and are sloughed (lost) from the tips of the villi daily and replaced by cells that arise from the Crypts of Leiberkuhn (nests of cells at the base of the villi—a rapidly dividing, undifferentiated stem cell population). Radiation damage in the small intestine is a result of damage to these cells.

Moderate doses of radiation result in shortening of the villi due to a decrease in mitotic activity of the crypt cells, followed by regeneration of these cells with a corresponding repopulation and healing of the villi. After high doses, the response is more dramatic, minimal recovery occurs, the villi become shortened and flattened and the intestine may become denuded (complete loss of cells) leading to ulceration, hemorrhage, fibrosis and necrosis (Fig. 5-7).

Changes in and damage to the large intestine and rectum after high doses are similar to those already outlined; the rectum, along with the esophagus, appears to be much more resistant to radiation than the stomach or small intestine.

DIAGNOSTIC RADIOLOGY AND NUCLEAR MEDICINE.—Neither diagnostic radiography and fluoroscopy nor radionuclide procedures deliver doses of a magnitude great enough to result in the type of changes detailed above.

RADIATION THERAPY.—Changes in the digestive system have implications in radiotherapy. Esophagitis and mucositis commonly occur during and posttreatment at doses of 1000 to 2000 rads but minimal

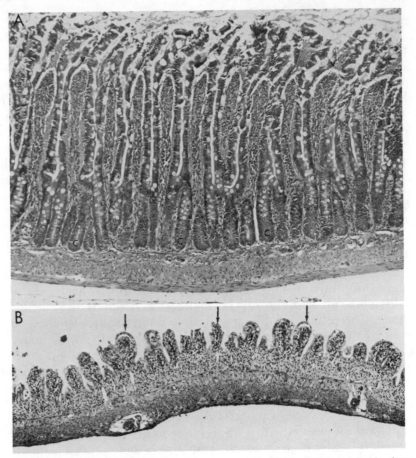

Fig. 5-7.—Small intestine of a rat exposed to 2000 rads total body dose, sacrificed 5 days postirradiation. **A,** normal intestinal mucosa with typical long villi *(V)*, and numerous crypts *(C)*. **B,** edema and blunting of the villi, loss of crypts, and almost total denudation of the intestinal mucosa *(arrows)*. (H & E stain, magnification × 40.)

late effects are observed even at total doses of 6000 to 7000 rads. Irradiation of the small intestine is often unavoidable when treating certain diseases (e.g., ovarian cancer by moving strip or whole abdomen technic), but late effects are minimal. Early effects are often manifested by symptoms such as nausea, vomiting and diarrhea. The incidence and severity of late changes in different parts of the GI tract increase with the total fractionated dose reflecting the sensitivity of the individual organs. For example, 9% of the patients receiving 5000 rads to the colon develop partial obstruction, while at doses greater than 6000 rads, 25% develop this late change.

Reproductive System

Male

Most of the tissues of the male reproductive system, with the exception of the testes, are radioresistant. The testes contain both nondividing, differentiated, radioresistant cells (mature spermatozoa) and rapidly dividing, undifferentiated, radiosensitive cells (immature spermatogonia). It is this latter cell population in the testes that accounts for the radiosensitivity of the system.

The primary effect of radiation on the male reproductive system is damage and depopulation of the spermatogonia, eventually resulting in depletion of mature sperm in the testes, a process termed *maturation depletion* (Fig. 5-8). A variable period of fertility occurs after testicular irradiation, attributable to the radioresistance of the mature sperm, and is followed by temporary or permanent sterility, depending on the dose. Sterility is due to a loss of the immature spermatogonia which divide and replace the mature sperm lost from the testes. Permanent sterility can be produced by an acute dose in the moderate range (500 to 600 rads) while a dose of 250 rads results in temporary sterility (twelve months duration).

Another potential hazard of testicular irradiation is the production of chromosome aberrations that may be passed on to succeeding generations. The fertile period occurring postexposure does not exclude the possibility of chromosome damage in functional spermatozoa. Chromosome changes in the immature spermatogonia also cannot be discounted and may either be propagated or eliminated through the many divisions needed to produce spermatozoa.

DIAGNOSTIC RADIOLOGY AND NUCLEAR MEDICINE.—These two clinical specialties involving acute low doses to patients and chronic low doses to personnel present no hazard in terms of sterility. These doses, however, may produce chromosomal changes possibly resulting in mutations in future generations. For this reason the utmost care should be taken to shield the testes from unnecessary radiation of all types.

RADIATION THERAPY.—In contrast to diagnostic procedures, total doses administered in radiotherapy can produce sterility in addition to chromosomal changes. In fact, it now appears that certain fractionation schedules produce sterility at *lower* total doses than a single exposure—the opposite of the expected. Every effort should be made to shield the testes from scattered radiation when the treatment field is situated close to this tissue. In addition, the patient should be informed of the possibility of temporary or permanent sterility and be given procreation advice when the situation warrants it. Another point

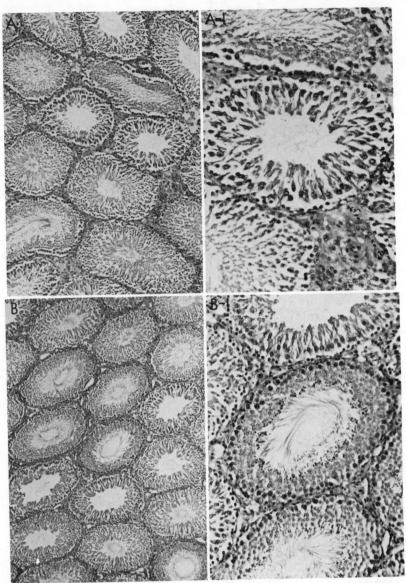

Fig. 5-8.—Testes from a rat exposed to 500 rads total body radiation. **A** and **A-1**, normal testes illustrating both immature and mature cell types. **B** and **B-1**, one week postexposure illustrating minimal changes; mature sperm are still present. **C** and **C-1**, three weeks; and **D** and **D-1**, four weeks postexposure, both times exhibit disappearance of normal architecture, looseness in the appearance of the network and a decrease in all cells, both precursor and mature, with a loss of polarity in the remaining cells. (H & E stain; A to D, magnification × 100; A-1 to D-1, magnification × 250.)

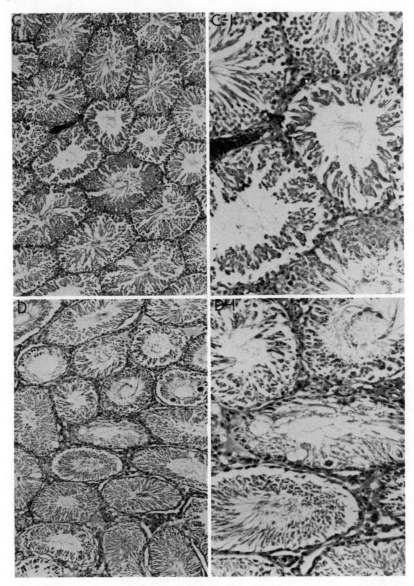

which must be clarified is that impotency is not caused by radiation-induced sterility.

Female

The ova are contained within sac-like enclosures (follicles), designated by size as small, intermediate and large, which vary in radiosensitivity as follows: intermediate follicles—most sensitive; small follicles—most resistant; and mature follicles—moderately sensitive. Unlike the male, these cells are not constantly dividing and replacing those lost through menstruation. An ovum is released from a mature follicle at ovulation, followed by either fertilization or, if this does not occur, menstruation. An initial period of fertility occurs after moderate doses of radiation to the ovaries, due to the presence of moderately resistant, mature follicles that can release an ovum. This fertile period is followed by temporary or permanent sterility resulting from damage to ova in the radiosensitive intermediate follicles, therefore inhibiting their maturation and release. Fertility may recur due to the maturation of ova in the radioresistant small follicles.

Although the dose necessary to produce sterility in females is a function of age (a higher dose is necessary in young women than in older women), in general a dose of greater than 625 rads produces sterility in women.

Major concern arises over the possibility of genetic changes in functional ova after irradiation. Although fertility recurs after low and some moderate doses, the possibility cannot be excluded that these functional ova have not incurred chromosome damage that may result in either grossly abnormal offspring or in offspring carrying nonvisible mutations that can be passed on to succeeding generations.

DIAGNOSTIC RADIOLOGY AND NUCLEAR MEDICINE.—As in the male, low doses received from diagnostic procedures do not produce sterility in the female but may cause chromosome changes. Although the body forms a natural shield for the ovaries, precautions always should be taken to avoid unnecessary ovarian exposure whenever possible.

RADIATION THERAPY.—Doses in the therapeutic range pose a double hazard—chromosomal damage and sterility. In addition, unlike the male where radiation-induced sterility does not produce effects in secondary sexual organs resulting in impotency, radiation sterilization of the female may produce an artificial menopause with marked effects on secondary genitalia and sexual characteristics. In those malignant diseases requiring irradiation of the ovaries, the patient should be informed of the possible consequences. Also, when sterility is not induced but the ovaries receive significant scattered radiation, procreation advice always should be given to the patient.

Eye

The lens of the eye contains a population of actively dividing cells that may be damaged and destroyed by radiation. Because there is no mechanism for removal of injured cells, those damaged cells form a cataract (i.e., a lens opacity). Moderate doses (as low as 200 rads) produce cataracts in a few individuals with the incidence increasing to 100% in individuals exposed to an acute dose of 700 rads. The degree of opacity is reflected as visual impairment, ranging from minimal impairment at 200 rads, becoming progressive and causing complete visual obstruction at higher doses. The frequency of cataracts varies with exposure to chronic and acute doses, with chronic doses producing a lower frequency of cataracts than acute doses.

DIAGNOSTIC RADIOLOGY.—Because radiation is scattered to the eye during fluoroscopic procedures, occupationally exposed personnel may exhibit cataracts. Particularly true in the early days of radiology when little knowledge existed of the biologic effects and hazards of radiation, cataracts are considered a late effect of radiation, appearing from 1 to 30 years postexposure (Chapter 7). Although doses in diagnostic radiology today are much lower and equipment is vastly improved, precautions should be taken to shield the eyes during fluoroscopic procedures.

Because of the high skin doses with dental x-rays, cataracts are a concern in dental radiography.

RADIATION THERAPY.—Doses to the eye in the therapeutic range certainly can induce cataracts; the minimal cataract inducing dose appears to be a total of 400 rads. Total doses of 1200 rads delivered by fractionated schedules induce cataracts in almost all patients, becoming progressive at 1400 rads. When treating lesions of the face located near the eyes, eye shields should always be used to reduce the formation and severity of cataracts.

Cardiovascular System

Vasculature

Blood vessels contribute to radiation damage in both radiosensitive and radioresistant organs; their response is particularly important in the latter. Vessel damage may result in occlusion through two possible ways: (1) damaged endothelial cells, or substances released from them, may stimulate division of undamaged cells in regenerative efforts (if too many cells are replaced, occlusion may occur) or (2) destruction of the endothelial cells may induce the formation of blood clots in the vessels (thrombosis). Small vessels, possibly due to their small lumens, are more radiosensitive than large vessels. These changes in blood vessels may be manifested in late changes such as petechial hemorrhages (pinpoint hemorrhages), telangiectasia (dilation of small terminal vessels) and vessel sclerosis (a type of fibrosis—actually a hardening and concomitant loss of elasticity of the vessel wall).

Because blood vessels are responsible for the transport of oxygen and nutrients to all organs of the body, occlusion of the vessels can have serious consequences to any organ. A complete loss of oxygen and nutrients will result in necrosis of the cells and tissues of that organ. A partial loss of these substances may result in atrophy and fibrosis of the organ, with a corresponding decrease in functional ability and a generalized decreased ability to withstand trauma.

Heart

For many years believed to be radioresistant, closer appraisal of the response of the heart to radiation is throwing doubts on this thought. Although undamaged at low and moderate doses except for functional (EKG) changes, high doses can produce pericarditis (inflammation of the pericardium—the membrane covering the heart) and pancarditis (inflammation of the entire heart).

DIAGNOSTIC RADIOLOGY AND NUCLEAR MEDICINE.—Doses in these two clinical areas are not of sufficient magnitude to produce changes in the heart.

RADIATION THERAPY.—The heart is often totally or partially included in the radiation field, for example, in the treatment of malignant lymphomas and of the chest wall following mastectomy for breast cancer. Fractionated doses totaling 4000 rads produce the above changes in a small percentage of individuals with the incidence increasing with increasing dose. When possible, the heart should be shielded during the entire treatment course and particularly when total doses exceed 4000 rads.

Growing Bone and Cartilage

Although mature bone and cartilage are radioresistant, growing bone and cartilage are moderately radiosensitive. In general, mature bone is formed through the mineralization (calcium deposition) of cartilage. Growing bone and cartilage consist of both nondividing, differentiated cells (osteocytes, chondrocytes) and rapidly dividing, undifferentiated cells (osteoblasts, chondroblasts); the latter group accounts for the moderate sensitivity of these tissues.

Damage to both small blood vessels and bone marrow also plays an important contributing role in radiation injury to growing bone. Moderate doses of radiation produce temporary inhibition of mitosis and death of the proliferating immature cells. Recovery does occur at doses of this magnitude resulting in minimal residual (late) damage.

High doses may produce a permanent inhibition of mitosis and destruction of proliferating cells resulting in cessation of bone formation. Few early gross changes are evident in bone even at these high doses. However, alterations in shape and size of the bone and scoliosis are evident late changes.

DIAGNOSTIC RADIOLOGY AND NUCLEAR MEDICINE.—The low doses administered during these procedures do not result in the changes outlined above. However, the administration of bone-seeking radionuclides, if not removed, can cause these changes.

RADIATION THERAPY.—Treatment of Wilm's tumor and neuroblastoma in children often unavoidably includes a growing bone in the radiation field, therefore producing bone abnormalities and scoliosis. Fractionated total doses greater than 2000 rads produce marked changes in the bones of children irradiated when less than two years of age. The incidence of bone abnormalities decreases with decreasing dose and increasing age at time of treatment.

Liver

The liver is usually considered part of the digestive system (an accessory gland essential to storage, metabolism and excretion of the products of the digestive process). For many years the liver was considered to be radioresistant, but current opinion is that the liver is a moderately sensitive, responsive organ.

The hepatic cells (parenchymal cells of the liver), while relatively resistant to radiation, do retain the capability of regeneration through mitosis. However, the liver has a large blood supply and a great number of both large and small blood vessels; for this reason, radiation injury to the hepatic cells is believed to be secondary to vascular changes.

Low and moderate doses produce little observable early response; early changes after high doses are difficult to detect except possibly through function studies. In some cases, the liver may be enlarged and fluid may accumulate in the abdominal cavity (ascites).

Delayed radiation effects in the liver, termed *radiation hepatitis*, are a consequence of vascular sclerosis and consist primarily of fibrosis (sometimes called *cirrhosis*) and possibly necrosis. The function of the liver is impaired, resulting in liver failure and jaundice.

DIAGNOSTIC RADIOLOGY AND NUCLEAR MEDICINE.—No observable response is detected in the liver subsequent to the doses delivered by these procedures.

RADIATION THERAPY.—The liver, either totally or partially, is sometimes included in the treatment field, e.g., the treatment of malignant diseases of the kidney, lymphomas and ovarian cancer with moving strip or whole abdomen technic. Doses in the clinical range produce radiation hepatitis, ranging from 3500 to 4500 rads total dose from a standard fractionation schedule. The clinical significance of this will depend on the volume of the organ irradiated.

Respiratory System

The respiratory system consists of the nose, pharynx, trachea and lungs. Although considered to be relatively resistant to radiation, the lungs are actually responsive to radiation in the high dose range (greater than 1000 rads).

The primary early change in the lungs after irradiation is inflammation, termed *radiation pneumonitis* (similar to pneumonia). This is a transitory response after moderate doses, and recovery occurs with minimal damage. A high dose to both lungs produces a progressive reaction that may develop from an early pneumonitis to late fibrosis, an outcome that can certainly cause death (Fig. 5-9).

DIAGNOSTIC RADIOLOGY AND NUCLEAR MEDICINE.—Radiation pneumonitis is not a response observed in the lungs after low doses.

RADIATION THERAPY.—One lung is often the primary treatment area, or a portion of it may be included in the treatment field of other organs, e.g., irradiation of the breast. Doses of 2500 rads to both lungs with a standard fractionation schedule may produce a progressive fibrosis in a small percentage (8%) of individuals treated. Increasing dose causes a corresponding increase in the number of patients with this response, reaching 50% at 3000 rads.

The response is dependent on the volume irradiated; one lung can be given a higher dose than both lungs. Although fibrosis occurs in the irradiated lung rendering it nonfunctional, the remaining undamaged lung continues to function.

Urinary System

The kidneys, ureters, bladder and urethra comprise the urinary system. The kidneys have the same relative sensitivity as the lungs and, like the lungs, radiation injury (termed *radiation nephritis*) appears to be a result of vascular damage. High doses produce few evident early changes other than a slight edema and swelling of the kidneys with late changes such as atrophy and fibrosis of the kidneys occurring secondary to blood vessel damage (particularly sclerosis). These changes result in hypertension (elevated blood pressure) and renal failure in the individual.

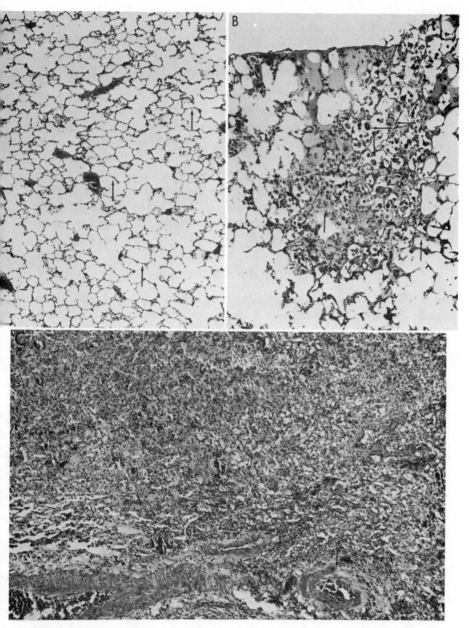

Fig. 5-9.—Rat exposed to an acute dose of 2000 rads to one lung only. **A,** normal lung showing honeycomb appearance with thin alveolar walls *(arrows)*. **B,** six months postirradiation, atypical changes in cells of alveoli *(arrow)*, thickened alveolar walls *(arrow)* and edema in the alveolar spaces *(E)*. **C,** one year postirradiation, total absence of lung parenchyma, now consisting of connective tissue and chronic inflammatory cell infiltrate, blood vessel *(arrow)* demonstrating radiation changes consisting of swollen endothelial (lining) cells and medial (middle layer) thickening. (H & E stain, magnification × 180.)

DIAGNOSTIC RADIOLOGY AND NUCLEAR MEDICINE.—Doses from diagnostic radiography, fluoroscopy and nuclear medicine do not produce this response in the urinary system.

RADIATION THERAPY.—When both kidneys are included in the treatment field, they must be shielded at a total dose of 2600 rads. The statistical incidence of these changes increases sharply after this dose —2800 rads in five weeks to both kidneys results in a high probability of fatal radiation nephritis.

As in the lungs, the volume irradiated plays a particularly important role. Exclusion of one-third of the kidney volume from the treatment field greatly minimizes renal failure. If occurring to only one kidney, the unirradiated kidney will continue to function if the irradiated one is surgically resected.

Central Nervous System

The nervous system consists of the brain and spinal cord. In general, the cells of the various parts of the nervous system are nondividing differentiated cells, rendering them relatively radioresistant. In fact, the nervous system is considered the most radioresistant system in the adult; therefore low and moderate doses of radiation will result in minimal, if any, morphologic damage (however, some authors have reported functional changes at low doses).

Early changes in the CNS after high doses include inflammation (termed *myelitis* in the spinal cord), progressing to necrosis and fibrosis of the brain or spinal cord as a result of blood vessel sclerosis and thrombosis. As in other radioresistant organs, radiation injury in the CNS appears to be a consequence of progressive vascular changes. Of particular interest is the comparatively higher radiosensitivity of the white than the gray matter of the brain. The threshold level for radiation injury to the CNS is between 2000 and 4000 rads.

DIAGNOSTIC RADIOLOGY AND NUCLEAR MEDICINE.—No observable changes result from doses of this low magnitude.

RADIATION THERAPY.—Doses in the clinical range totaling 5000 rads can cause delayed radiation necrosis in the brain. Because brain irradiation occurs for treatment of brain tumors, it is often difficult to distinguish the effects caused by the tumor from those caused by irradiation. The response of the spinal cord varies with the volume and area irradiated; cervical and thoracic cord are both more sensitive than lumbar. The incidence of radiation myelitis increases at doses greater than 5000 rads given to small volumes and greater than 4500 rads to large volumes. This is of importance because the cord is often included in the treatment field of many diseases, e.g., lung cancer, cancer of the esophagus, Hodgkin's disease and tumors of the head and neck region.

Functional Response

The discussion of radiation response of tissues and organs in this chapter has centered on morphologic (i.e., structural) changes, with functional changes discussed in relation to these. Admittedly, structural alterations in an organ, particularly if dramatic (such as necrosis or fibrosis), result in a corresponding loss of function. However, the function of an organ or cell may be altered before structural changes are evident, that is, functional changes may appear at an earlier time or at a lower dose.

In comparison to the number of studies concerning morphologic changes, observations of functional response are minimal. Studies observing organ function following irradiation indicate that functional changes occur at lower doses than do morphologic changes—appearing to be particuarly true in morphologically radioresistant organs and systems such as the CNS. Doses far below those causing morphologic response, on the order of 10 rads, have been reported as causing functional changes in the CNS. This same situation may be true in terms of other morphologically radioresistant organs. In essence, we may be dealing with radioresistant organs in terms of morphology but with radiosensitive organs in terms of function. Further investigation of functional response of organs to radiation is necessary to define both these changes and their impact on the health of the individual.

Response to Chronic Low-Dose Radiation

The response of tissues and organs to chronic low-dose radiation is another topic that requires further study. This area is becoming increasingly important with the increased use of radiation—particularly in terms of occupationally exposed individuals. The major concern from chronic low-dose radiation is the possibility of genetic damage in these cells or the occurrence of so-called "late effects" of radiation such as cancer.

REFERENCES

1. Anderson, W. A. D.: *Pathology* (6th ed.; St. Louis: C. V. Mosby, 1971).
2. Berdjis, C. C.: *Pathology of Irradiation* (Baltimore: Williams & Wilkins, 1971).
3. Casarett, A. P.: *Radiation Biology* (Englewood Cliffs, New Jersey: Prentice-Hall, 1968).

4. Dalrymple, G. V., *et al.*: *Medical Radiation Biology* (Philadelphia: W. B. Saunders, 1973).
5. Fletcher, G. H.: *Textbook of Radiotherapy* (2nd ed.; Philadelphia: Lea & Febiger, 1973).
6. Moss, W., *et al.*: *Radiation Oncology: Rationale, Technique, Results* (4th ed.; St. Louis: C. V. Mosby, 1973).
7. Robbins, S. L.: *Pathologic Basis of Disease* (Philadelphia: W. B. Saunders, 1974).
8. Rubin, P., and Casarett, G. W.: *Clinical Radiation Pathology*, Vols. I and II (Philadelphia: W. B. Saunders, 1968).

6/Total Body Radiation Response

The response of the organism to total body radiation is determined by the combined response of all systems in the body. Because systems differ in their radiosensitivity and therefore response, the total body response will be a function of the particular system most affected by the radiation. The previous chapter dealt with the response of specific organs and systems to radiation; these findings will now be applied to the response of the organism to acute total body radiation, in both adult and fetal life.

Adult

The response of an adult organism to an acute total body exposure to radiation results in specific signs, symptoms and clinical findings. The relationship of these signs and symptoms to a specific type of trauma or disease process is termed a *syndrome*. Because the response of the organism to total body radiation results in specific findings, the term "total body syndrome" or "radiation syndrome" is used. Although damage to one particular system is responsible for the syndrome, the manifestation of the specific signs and symptoms are a result of damage in more than one system in the body.

Radiation Syndrome in Mammals

The term "total body radiation syndrome" applies only when exposure occurs under a specific set of conditions. First, exposure of the organism must have occurred acutely—to be more exact—in a matter of minutes rather than hours or days. Second, the area of the organism exposed to radiation must be total body, or very nearly total body, to manifest the full-blown radiation syndrome. Third, the radiation syndromes are produced by exposure to external penetrating sources such as x-rays, γ-rays and neutrons. Radioactive materials deposited internally do not result in the appearance of the full syndrome.

SURVIVAL TIME.—The primary effect of an acute total body exposure is to shorten the lifespan of the organism; the time of lifespan shortening is dependent on the dose to which an organism is exposed. Because the lifespan of an organism is drastically reduced after a

109

moderate to high dose of total body radiation (the organism may live as long as 1 to 2 months postexposure or only hours postexposure), total body exposures in this dose range are considered immediately lethal.

The survival time of an organism exposed to total body radiation is expressed as the mean (average) survival time. Variations in survival time exist between different species and even between animals within the same species. Some animals live longer than others due to these individual variations in sensitivity. The expression of survival time as the mean takes these variations into account, therefore eliminating those individuals who appear to be extremely radiosensitive or radioresistant. In this way a general pattern of survival time will appear when large numbers of animals are irradiated.

Because of the variations in survival time of a group of animals exposed to the same total body dose, the relationship of survival time of an entire population of the same species (e.g., all dogs or all monkeys) exposed to the same total body dose is expressed as that dose which kills a certain percentage of the population within a given period of time. For example, the lethal dose necessary to kill 50% of a population in 30 days is expressed as the $LD_{50/30}$ dose; the lethal dose necessary to kill 100% of the population in 6 days is the $LD_{100/6}$ dose. The $LD_{50/30}$ dose is the most often-used expression when discussing total body exposures although in humans the expression $LD_{50/60}$ dose is gaining more widespread use. The reasons for this will be discussed in a later section. Table 6-1 lists the $LD_{50/30}$ doses in rads for different species.

DOSE AND SURVIVAL TIME.—A curve can be constructed relating survival time to dose for all mammals exposed to total body radiation

TABLE 6-1.—$LD_{50/30}$ VALUES FOR DIFFERENT SPECIES

SPECIES	$LD_{50/30}$* (RADS)
Human	250–300
Monkey	400
Dog	300
Rabbit	800
Rat	900
Mouse	900
Chicken	600
Frog	700
Goldfish	2000

*In humans the expression $LD_{50/60}$ may be more useful (for explanation, see the text). In addition, this figure is an estimate based on the small number of accidental overexposures; more data may change these estimates.

(Fig. 6-1). As dose increases, the number of survivors and survival time decrease accordingly. At a dose of approximately 200 R, death will occur in a small percentage of the animals, this percentage increasing with increasing dose. In this dose range, survival time is dose-dependent, i.e., decreasing with increasing dose. Mean survival time does not appear to be a function of dose between 1000 and 10,000 R. All animals irradiated with a total body dose in this range survive approximately the same time; fewer animals survive these doses than the lower doses. Mean survival time becomes dose-dependent (decreases with increasing dose) at doses of 10,000 R and over; in addition, the number of survivors decreases.

The three general regions of the curve in Figure 6-1 reflect damage to three different systems that may result in the death of the animal. This does not indicate that other organs and systems have not sustained radiation damage—but the primary cause of death is destruction of one specific system. The three defined radiation syndromes are named according to failure of that organ system which is responsible for death.

Between 100 and 1000 R, death occurs primarily as a result of damage to the hemopoietic system; in particular, to destruction of the bone marrow. The total body syndrome reflected in this dose range is termed the *hemopoietic* or, more appropriately, the *bone marrow syndrome*.

Fig. 6-1.—Curve derived from studies of the response of different species to total body irradiation. Note the three distinct variations in survival time in the dose ranges 200–1000 R, 1000–10,000 R and >10,000 R, reflecting damage to the specific organ system for which the radiation syndromes are named. (From Pizzarello, D. J., and Witcofski, R. L.: *Basic Radiation Biology* [Philadelphia: Lea & Febiger, 1967]. Courtesy of the publisher.)

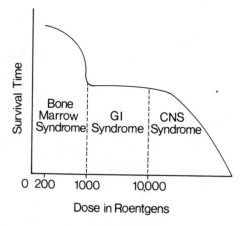

In the second region of the dose response curve (1000 to 10,000 R), death is primarily due to damage in the gastrointestinal system, particularly in the small intestine. The syndrome reflected in this dose range is the *gastrointestinal (GI) syndrome*.

Doses greater than 10,000 R comprise the third area of the curve and reflect damage in the central nervous system. Because damage in this system is responsible for the death of the animal, the syndrome is referred to as the *central nervous system (CNS) syndrome*.

Two points must be kept in mind concerning the generalized curve in Figure 6-1. First, the dose ranges given for each syndrome are not specific for humans; these figures are derived from studies of the response of many different animals following acute total body exposure. Second, there is an overlap of the syndromes at the higher doses of each dose range, e.g., between 600 and 1000 R, some animals die from a combination of damage in the hemopoietic *and* gastrointestinal systems. Therefore, the *threshold* for the GI syndrome may be 600 R. However, because the majority of animals die primarily from destruction of the GI tract at a dose of 1000 R, the GI syndrome is defined between 1000 and 10,000 R. This same reasoning applies to the CNS syndrome. Doses less than 10,000 R will cause death from CNS damage in some animals, but the majority of animals die primarily from CNS destruction at doses greater than 10,000 R. Therefore, the threshold for the CNS syndrome is lower than 10,000 R in spite of the fact that the syndrome is defined at this dose.

STAGES OF RESPONSE.—The response of an animal to an acute total body dose of radiation can be divided into three stages. The length of time that each stage persists is dose-dependent, i.e., the lower the dose, the longer the duration of any of the three stages. Each stage may last from a period of weeks (low doses) to a matter of minutes (high doses).

The first phase is the *prodromal stage* which is characterized by nausea, vomiting and diarrhea (often referred to as the N-V-D syndrome). The prodromal stage may last from a few minutes to a few days, depending on the dose (i.e., the higher the dose, the shorter the prodromal stage).

The second phase is the *latent stage*; the term is derived from the generally well appearance of the animal during this time. However, changes are taking place in the respective systems damaged by the radiation that will eventually lead to either the demise or recovery of the animal. The length of time of the latent stage varies with dose (i.e., from weeks at doses below 500 R, to hours or less at doses greater than 10,000 R).

After this time the animal becomes obviously ill and exhibits the specific signs and symptoms of the particular syndrome reflecting the organ system damaged. This third phase is termed, appropriately, the *manifest illness stage* and also will last from minutes to weeks depending on the dose. Finally, the animal either recovers or dies as a result of radiation injury. Table 6-2 summarizes these stages.

The three radiation syndromes will be discussed in relation to findings in humans. Although the time of appearance of the signs and symptoms varies among different mammals, generally speaking, all mammals will exhibit similar signs and symptoms. Survival time following different doses varies greatly among species, with humans appearing to be a relatively radioresponsive species (see Table 6-1).

Radiation Syndrome in Humans

Data concerning the exposure of humans to acute total body radiation has been accumulated from the following sources:

1. Accidents in industry and laboratories (approximately 50 reported cases).
2. Pacific Testing Grounds accidents involving exposure to fallout.
3. The exposure of individuals at Hiroshima and Nagasaki.
4. Medical exposures involving total body or near total body radiation for cancer therapy or other reasons.

Although all of these situations provide an opportunity to study the response of humans to total body radiation, difficulties arise over the estimation of dose and the extent of exposure received by individuals in all cases except possibly one—medical exposure. In spite of these problems, the study of these individuals has resulted in a general pattern of the radiation syndromes in humans. One important point to keep in mind is that in most of these individuals medical sup-

TABLE 6-2.—STAGES OF THE TOTAL BODY SYNDROME

STAGE	TIME	MAJOR SYMPTOMS
Prodromal	Days–minutes	N-V-D syndrome Nausea Vomiting Diarrhea
Latent	Weeks–hours	None
Manifest illness	Weeks–hours	Symptoms reflect the systems damaged *Bone Marrow*—fever, malaise *Gastrointestinal*—malaise, anorexia, severe diarrhea, fever, dehydration, electrolyte imbalance *CNS*—lethargy, tremors, convulsions, nervousness, watery diarrhea, coma

port resulted in an increased survival time. Without this life-sustaining support, survival time would be decreased.

BONE MARROW SYNDROME.—The bone marrow syndrome in humans occurs between 100 and 1000 R with death occurring in a few individuals at a dose of 200 R. The $LD_{50/60}$* for humans is approximately 450 R or 250–300 R absorbed dose (rads) and is in the dose range of the bone marrow syndrome. Death of the individual in this syndrome is due to destruction of the bone marrow to an extent that will not support life.

The prodromal stage of the bone marrow syndrome occurs a few hours postexposure and consists mainly of nausea and vomiting. The latent stage lasts from a few days to 3 weeks postexposure, during which time the number of cells in the circulating blood are not severely depressed. The individual appears and feels well during this time; however, stem cells in the bone marrow are dying during both the prodromal and the latent stages resulting in a decreased production of mature cells and, therefore, a decrease in the number of cells in the circulating blood. This drop in blood cell count appears during the manifest illness stage, occurring at 3 weeks and possibly continuing through the fifth week postexposure (i.e., at low doses). During this time the individual exhibits the specific signs and symptoms of the bone marrow syndrome. The depression of all blood cell counts (pancytopenia) results in anemia (due to depression of RBCs and hemorrhage) and serious infections (due to the depression of white cells—leukopenia).

Survival in the dose range of the bone marrow syndrome decreases with increasing dose. At the lower limits of the dose range, 100–300 R, the bone marrow will become repopulated to an extent great enough to support life in the majority of individuals (Fig. 6-2). Full recovery of a large percentage of these individuals will occur from 3 weeks to 6 months postexposure. A few sensitive individuals may die 6 to 8 weeks after an exposure of 200 R; doses from 400 to 600 R (dose range of the $LD_{50/60}$) result in a decreased number of survivors. In these individuals partial repopulation of the bone marrow may occur, but not to a sufficient extent to support life. No human has been reported as surviving a dose of 1000 R.

As dose increases, survival time decreases—death occurring in

_____

*Because humans appear to both develop and recover from the bone marrow syndrome later than other species, deaths may occur as late as 60 days postexposure. In addition, the majority of deaths occur by 30 days in humans whereas in animals they occur by 15 days. For these reasons the expression $LD_{50/60}$ may be more meaningful in humans than the $LD_{50/30}$ used in animals.

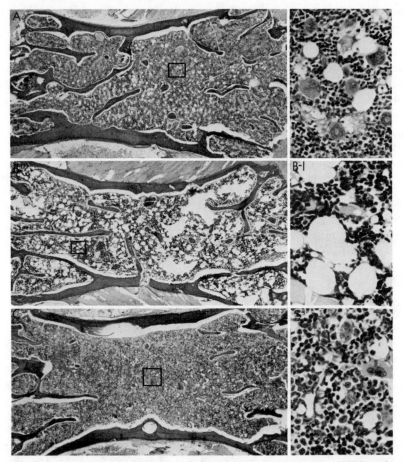

Fig. 6-2.—Rat bone marrow following a total body dose of 500 rads. **A,** normal. **B,** one week postexposure exhibiting hypocellularity and relative increase in fat. **C,** three weeks postexposure, contents now appear relatively normal indicating regeneration (absence of bony trabeculae in C is not a result of irradiation but is due to the area of the tissue sectioned). (H & E stain; A to C, magnification × 40; A-1 to C-1, magnification × 300.)

approximately 4 to 6 weeks at doses of 300 to 500 R but in 2 weeks at doses of 500 to 1000 R.

The primary cause of death from the bone marrow syndrome is destruction of the bone marrow resulting in infection and hemorrhage. The bone marrow is normally filled with cells that supply mature cells to the circulating blood. After exposures in this dose range, the number of cells in the bone marrow steadily decreases until the bone marrow is not capable of producing those cells which are necessary in the circulating blood to sustain life (Fig. 6-3).

GASTROINTESTINAL (GI) SYNDROME.—The second defined acute radiation syndrome is the gastrointestinal (GI) syndrome. Doses between 1000 and 10,000 R will result in the GI syndrome in all animals studied; however, in humans some symptoms of the GI syndrome appear at a dose of 600 R (the threshold dose). The full syndrome is apparent at 1000 R. The LD_{100} for humans (between 600 and 1000 R) is

Fig. 6-3.—Rat bone marrow following a total body dose of 1000 rads (**B** and **B-1**) and 2000 rads (**C** and **C-1**), both illustrating a dramatic hypocellularity and increased fat content compared to normal (**A** and **A-1**). Those cells present are red blood cells due to hemorrhage. (H & E stain; A to C, magnification × 40; A-1 to C-1, magnification × 300.)

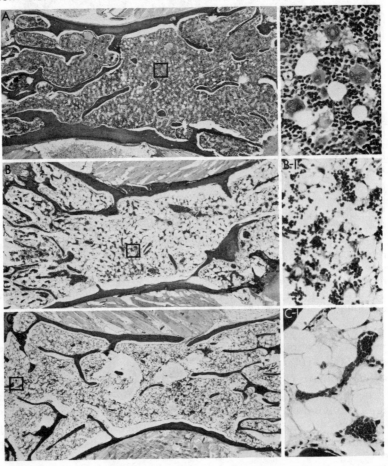

in the dose range of the GI syndrome. Survival time does not vary
with dose in this syndrome; death occurs at the same time regardless
of dose. In humans death occurs within 3 to 10 days if medical support
is not administered and within approximately 2 weeks even with medical support.

The prodromal stage of the GI syndrome occurs within a few
hours postexposure and is characterized by severe nausea and vomiting, which may be accompanied by severe cramps and diarrhea. From
the second through the fifth day, the individual enters the latent stage
and feels well. At the end of this time, there is a recurrence of severe
diarrhea, nausea and vomiting accompanied by fever, signaling the
onset of the manifest illness stage, which may persist from the fifth
through the tenth day. Death occurs from the GI syndrome during the
second week postexposure if life-sustaining support has been administered (fluids given, transfusions, etc.).

The GI syndrome is due to damage in two organ systems: the gastrointestinal tract and the bone marrow. The full GI syndrome does
not occur if only the GI tract has been irradiated because the bone
marrow plays an integral role in this syndrome.

The lining of the GI tract, particularly the small intestine, is severely damaged by doses in this range. The mitotic activity of the cells
in the Crypts of Leiberkuhn, the radiosensitive precursor cells to the
population of cells on the villi, will be decreased drastically following
exposure. As a result the villi, which slough dead cells into the intestinal lumen every 24 hours and are dependent on the Crypts of Leiberkuhn for replacements, lose cells and become shortened, flattened
and partially or completely denuded (Figs. 6-4 and 6-5).

The consequences to the individual of these changes in the GI
tract are profound. The flattened villi result in decreased absorption of
materials across the intestinal wall. Fluids leak into the lumen of the
GI tract resulting in dehydration. Overwhelming infection occurs as
bacteria that normally live within the GI tract gain access to the bloodstream through the intestinal wall causing systemic infection.

The effects of these drastic changes in the GI tract are compounded by equally drastic changes in the bone marrow. In fact, the
effects of damage in the bone marrow occur at a time when damage in
the GI tract is reaching its maximum. Of primary importance is the
severe decrease in the number of circulating white cells. This depression occurs as bacteria are invading the bloodstream from the GI tract,
therefore compounding an already severe problem. The remaining
blood cells may not exhibit a severe decrease in numbers due to the
fact that death occurs before radiation damage is reflected in these cell
lines.

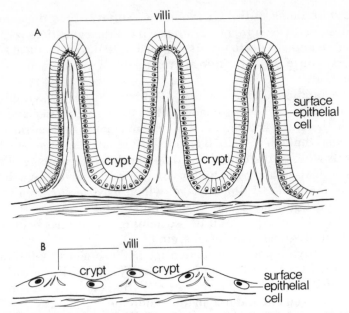

Fig. 6-4.—Diagrammatic representation of changes in the small intestine following total body exposure in the dose range of the GI syndrome. **A,** pre-irradiation; **B,** postirradiation.

Although attempts at regeneration occur in the GI tract after irradiation, particularly at the lower dose levels of the GI syndrome, the damage incurred by the bone marrow will probably still result in death (Fig. 6-6). Death from the GI syndrome is due primarily to infection, dehydration and electrolyte imbalance resulting from the destructive and irreparable changes in the GI tract and bone marrow.

CENTRAL NERVOUS SYSTEM (CNS) SYNDROME.—The full CNS syndrome occurs at a dose of greater than 5000 R in humans. Although CNS damage is evident at lower doses (2000 R), death from the full CNS syndrome occurs within 2 to 3 days following an exposure of 5000 R to all individuals.

The prodromal stage varies from a few minutes to a few hours dependent on dose. The signs and symptoms of this phase are extreme nervousness, confusion, severe nausea and vomiting, a loss of consciousness and complaints of burning sensations of the skin. A latent period next appears and may last for several hours, although often it is of shorter duration. The manifest illness stage begins 5 to 6 hours postexposure, at which time there is a return of watery diarrhea, convulsions, coma and finally death.

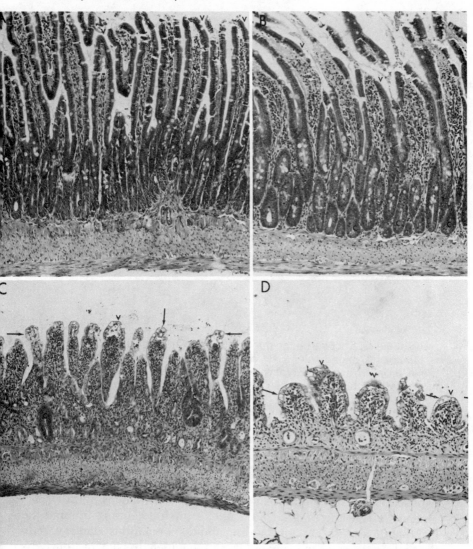

Fig. 6-5.—Small intestine of rats following various doses of total body irradiation, all sacrificed 5 days postexposure. **A,** normal, note villi *(V)* and crypts *(C)*. **B,** 500 rads, minimal shortening and sloughing of cells from villi is evident. **C,** 1000 rads, blunted villi with atypical epithelial cells *(arrows),* crypts contain mitotic figures. **D,** 2000 rads, exhibiting absence of crypts, sloughing and edema of villi. Note atypical changes in epithelial cells at this dose *(arrows).* (H & E stain, magnification × 75.)

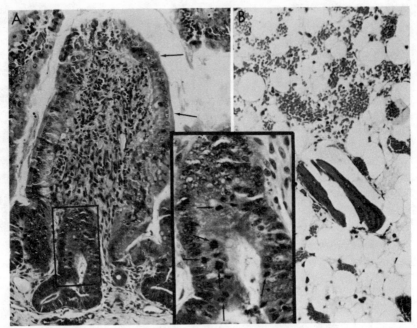

Fig. 6-6.—GI tract and bone marrow of rat exposed to 1000 rads total body irradiation. **A,** high-power view of edematous villi exhibiting atypical epithelial cells *(arrows)* and crypt with numerous mitotic figures. Insert indicates regenerative attempts *(arrows)*. **B,** bone marrow from same animal. Note the absence of all stem cells, only red blood cells are present. Although regenerative efforts occur in the crypts, the damage in the bone marrow is sufficient to cause death of the animal. (H & E stain; A and B, magnification × 175; insert, × 350.)

The cause of death in the CNS syndrome is not fully known or understood. Examination of the CNS after an individual has been exposed to a dose within this range reveals few changes in the parenchymal cells of the brain. This is not surprising keeping in mind the fact that these cells are nondividing and do not manifest damage as do the cells in the bone marrow or GI tract in which division of stem cells is necessary to provide the end cell to carry on the function of those systems. Damage in the CNS may be a result of damage to the blood vessels that supply the system resulting in edema in the cranial vault, vasculitis (inflammatory changes in the vessels) and meningitis (inflammation of the meninges). Death is suggested to be due to increased pressure in the confining cranial vault as a result of increased fluid content caused by these changes.

The bone marrow and the GI tract do not exhibit dramatic changes in the CNS syndrome because the individual does not live long enough. An increased survival time would result in the manifestation of dramatic changes in these systems.

The three radiation syndromes described above are outlined in Table 6-3.

Embryo and Fetus

Radiation has long been known to have profoundly damaging effects on the developing embryo. For obvious reasons, systematic studies of the effects of radiation on the developing embryo have been conducted in laboratory animals, particularly mice and rats. These studies have resulted in a wealth of information on this subject, including the definition of specific effects induced by radiation.

There are three general effects of radiation on the embryo and fetus:

1. Lethality.
2. Congenital abnormalities present at birth.
3. Long-term effects (late effects) that are not visible at birth but develop later in life.

These effects can be produced in the embryo and fetus by a mutation in the ovum or sperm resulting in inherited (genetic) effects, or they can be directly induced by exposure of the fetus to radiation (congenital).

This section will discuss those effects induced by irradiation in utero (congenital effects) and not those transmitted through a mutated ovum or sperm (genetic effects). In addition, the discussion will be limited to the lethal effects and congenital abnormalities induced by radiation that are present at birth or shortly thereafter. Late effects will be discussed in a subsequent chapter. Because laboratory animals have been the primary source of information concerning this topic, the effects discussed will be those observed in these animals, unless otherwise stated. The extrapolation and implication of these findings for humans and observations in humans are presented at the end of this chapter.

Fetal Development

Lethality and specific gross abnormalities induced in the embryo and fetus by radiation are dependent on the time of gestation (in fact, the *part of day* of gestation) at which exposure occurs. For this reason, a basic knowledge of fetal development will give the reader a better understanding of radiation effects.

Russell and Russell[18] have divided fetal development into three general stages: *pre-implantation, major organogenesis* and *fetal (growth) stage*. In humans the *pre-implantation stage* occurs from

TABLE 6-3.—SUMMARY OF ACUTE RADIATION SYNDROMES IN HUMANS AFTER WHOLE-BODY IRRADIATION*

SYNDROME	DOSE RANGE	TIME OF DEATH	ORGAN AND SYSTEM DAMAGED	SIGNS AND SYMPTOMS FINDINGS	RECOVERY TIME
Hemopoietic	100–1000 R†	3 weeks to 2 months	Bone marrow	Decreased number of stem cells in bone marrow, increased amount of fat in bone marrow, pancytopenia, anemia, hemorrhage, infection	Dose dependent—3 weeks to 6 months; some individuals do not survive
GI	1000–5000 R‡	3 to 10 days	Small intestine	Denudation of villi in small intestine, neutropenia, infection, bone marrow depression, electrolyte imbalance, watery diarrhea	None
CNS	> 5000 R	< 3 days	Brain	Vasculitis, edema, meningitis	None

*From Rubin, P., and Casarett, G. W.: *Clinical Radiation Pathology*, Vol. II (Philadelphia: W. B. Saunders, 1968).
†LD$_{50/60}$ for humans in this dose range (450 R).
‡LD$_{100}$ for humans in this dose range (1000 R).

conception to 10 days postconception and precedes implantation of the embryo in the uterine wall. During this time the fertilized ovum repeatedly divides forming a ball of cells which are highly undifferentiated.

Implantation of this ball of cells (the embryo) in the uterine wall signals the onset of the second stage—*major organogenesis*—which extends through the sixth week postconception in humans. During this time the cells of the embryo begin differentiating into the various stem cells that eventually will form all the organs of the body. The initial differentiation of cells to form a certain organ occurs on a specific gestational day. In humans, for example, neuroblasts (stem cells of the CNS) appear on the eighteenth gestational day, the forebrain and eyes begin to form on the twentieth day and on the twenty-first day the primitive germ cells appear.

At the end of the sixth week postconception, the embryo is termed a fetus and enters the *fetal stage*, primarily a period of growth. The fetus at this time contains most organ systems and many types of cells ranging from undifferentiated stem cells to more differentiated cells.

One specific system that should be given further attention is the development of the central nervous system (CNS). In adults, the CNS consists primarily of nondividing highly differentiated cells; the fetal CNS is in direct contrast to this. The neuroblasts appear at a very early point in fetal development (in humans on the eighteenth day) and are the most abundant and scattered cells present in all stages of embryonic development. As development progresses and the fetus grows in size, the neuroblasts become more diffusely dispersed throughout the body, in addition to undergoing some differentiation and becoming less mitotically active. However, the majority of these cells continue to exist throughout fetal development and until at least 2 weeks after birth. In fact, complete development of the CNS in humans may not occur until 10-12 years of age.

Based on the characteristics of fetal development and the fact that radiosensitivity is related to mitotic activity and differentiation, the fetus can be expected to be highly vulnerable not only to the lethal effects of radiation but also to the induction of gross abnormalities recognizable at birth. After conception, one cell—the fertilized ovum—repeatedly divides and differentiates producing the millions of various cells in the newborn animal. It is not surprising, then, that the developing embryo and fetus is exquisitely sensitive to radiation.

Radiation Effects on Mouse Fetal Development

PRE-IMPLANTATION.—The effects of radiation on the embryo and fetus are a function of the stage of development during which exposure occurs. Figure 6-7 illustrates the effects of 200 R on mouse embryos exposed on different gestation days. The early embryo during pre-implantation and major organogenesis is exquisitely sensitive to ionizing radiation. Exposure during the pre-implantation stage results in a high incidence of prenatal death (death of the embryo before birth). This is not surprising because the embryo consists of relatively few cells at this time, therefore damage to one cell, the progenitor of many descendant cells, has a high probability of being fatal. Doses as low as 10 R have been reported as being fatal to the mouse embryo during this time. Those embryos which do survive exhibit few congenital abnormalities at birth; one specific abnormality that has been reported as a result of irradiation during pre-implatation is exencephaly (brain hernia, or protrusion of the brain through the top of the skull).

MAJOR ORGANOGENESIS.—The incidence of congenital abnormalities in mouse embryos increases dramatically when exposure occurs during major organogenesis. Gross abnormalities have been observed in the mouse embryo exposed to 25 R during this stage. This is also the stage during which rubella (German measles) and the drug Thalidomide are believed to wreak havoc on the fetus.

Although all major organs are beginning to form during this time, differentiation of cells to form various organs begins on specific days. As a result, irradiation on certain gestational days will result in specific abnormalities. For example, exposure of the mouse embryo on the ninth day results in a high incidence of ear and nose abnormalities, whereas exposure on the tenth day results in bone abnormalities. The greatest variety of congenital abnormalities are produced when radiation is given during the eighth to the twelfth day in the mouse, corresponding to the twenty-third to the thirty-seventh day in humans. The time in which these abnormalities are produced is short, appearing to be when the cells initially differentiate and assume characteristics of the developing organ.

The majority of the effects of radiation on the fetus during this period of development are manifested in the CNS and related sense organs such as the eye. The sensitivity of the CNS is related to the abundance and immature characteristics of neuroblasts. Unlike the adult, where the majority of cells in the CNS are nondividing, highly differentiated and radioresistant, the neuroblasts in the fetus are highly undifferentiated, actively mitotic radiosensitive cells. These

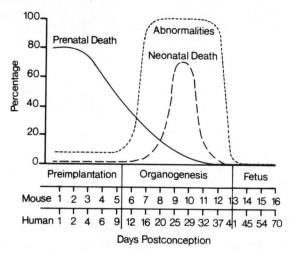

Fig. 6-7.—Relationship of 200 R in utero exposure on different gestational days to the induction of lethality and major abnormalities in the mouse embryo. Lower scale indicates Rugh's time estimates for the three stages in the human embryo. (From Russell, L. B., and Russell, W. L.: An analysis of the changing radiation response of the developing mouse embryo, J. Cell. Physiol. [Suppl. 43] 1:103, 1954. Courtesy of the Wistar Press.)

cells are scattered throughout the fetus resulting in a high incidence of abnormalities involving the neurologic system. Some of the most common abnormalities of the CNS observed in mice after in utero irradiation include brain abnormalities such as microcephaly (small brain), hydrocephaly (water on the brain) and eye deformities such as microphthalmia (small eyes). Behavioral abnormalities such as mental retardation have been reported in humans.

The developing musculoskeletal system also appears to be radiosensitive but not to the same degree as the CNS. Skeletal abnormalities such as stunted growth, abnormal limbs and others have been observed in the mouse fetus irradiated during the time of major organogenesis.

The incidence of prenatal death decreases when exposure occurs during major organogenesis; however, there is an increase in neonatal death (death at birth). This may be partially due to the presence of abnormalities in the fetus that are fatal at term. Table 6-4 lists some of the most common abnormalities observed in rodents and humans following in utero exposure during major organogenesis. Figures 6-8, 6-9 and 6-10 show animals with gross abnormalities resulting from in utero irradiation.

TABLE 6-4.—SOME MAJOR ABNORMALITIES FOUND IN MAMMALS
(HUMANS, RABBITS, MICE) AFTER FETAL IRRADIATION*

CNS	SKELETAL	OCULAR	OTHERS
Exencephaly	Stunting	Absence of eye(s)	Leukemia
Microcephaly	Abnormal limbs	Microphthalmia	Genital deformities
Mental retardation	Small head	(small eyes)	with sterility
Idiocy	Cleft palate	Strabismus	
Skull malformations	Club feet	Cataract	
Hydrocephaly	Deformed arms	Absence of lens	
Mongolism	Spina bifida		

*From Rugh, R.: The impact of ionizing radiation on the embryo and fetus, Am. J. Roentgenol. Radium Ther. Nucl. Med. 89:182, 1963.

FETAL (GROWTH) STAGE.—Irradiation during the fetal period results in fewer *obvious* abnormalities and a decreased incidence of both prenatal and neonatal deaths. Higher doses are necessary during this time to produce lethality and gross abnormalities. This is not surprising because the cells in the fetus are more differentiated than at earlier stages of development. However, irradiation during this period of gestation may result in effects that occur later in life (e.g., cancer) or in functional disorders after birth.

Fig. 6-8.—Two rats of the same litter exposed to x-rays in utero. **A,** rat exhibiting almost total anophthalmia. **B,** rat exhibiting normal right eye but a degree of anophthalmia in the other eye. (Courtesy of Dr. Roberts Rugh.)

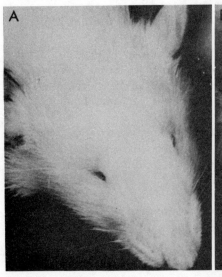

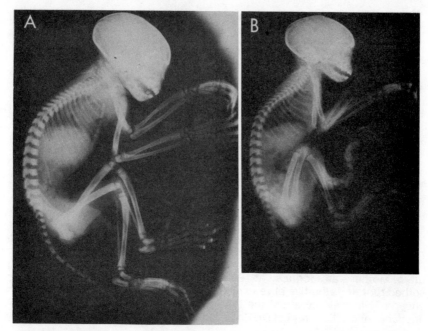

Fig. 6-9.—Radiograph of two-week-old monkey. **A,** control. **B,** animal exposed to x-rays at 13 days of gestation, indicating stunting. At 13 days, the skeletal elements are differentiating. (Courtesy of Dr. Roberts Rugh.)

Radiation Effects on Human Embryos

The devastating effects of radiation on the developing human embryo and fetus have been a subject of major concern for many years and are of particular interest today with the increased use of ionizing radiation for medical purposes. This is also a very controversial subject in terms of abnormalities produced by clinical doses, especially in the diagnostic range. Many reports have appeared in the literature implicating radiation as the cause of a specific anomaly. Although it is well known that radiation does have a very dramatic effect on the fetus in terms of both lethality and the induction of congenital abnormalities, it is difficult to establish a causal relationship between radiation and a specific abnormality. Two reasons for this are as follows:

1. The incidence of spontaneous congenital abnormalities in the population is approximately 6%.
2. Radiation induces no *unique* congenital abnormalities (i.e., radiation-induced congenital abnormalities are the same as those that appear spontaneously or those caused by other factors).

These two factors make it difficult to implicate radiation as the sole cause of a specific congenital abnormality.

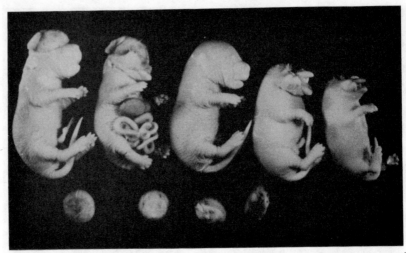

Fig. 6-10.—Litter of rats exhibiting anomalies after exposure to x-rays in utero. Mother was sacrificed at 19 days postexposure. Four fetuses were resorbed *(bottom);* the five alive exhibited, from left to right, exencephaly, exencephaly and evisceration, the third appears normal and the last two are anencephalics. (Courtesy of Dr. Roberts Rugh.)

For obvious reasons, systematic studies of the effects of radiation on fetal development have been derived from laboratory animals, particularly mice. Although it is generally accepted that abnormalities produced in the mouse fetus by radiation also can be produced in humans, there are two factors that must be considered when extrapolating these findings to humans. One factor is time; the gestation period in mice is 20 days—in humans it is 270 days. Because the induction of specific anomalies is dependent on the period of development (i.e., gestation day) during which irradiation occurs, there will be a difference in the time these effects are induced in humans as compared to rodents.

The second important factor to consider is dose. The question arises as to whether the human embryo is more or less sensitive than the rodent embryo. Comparative studies between mice and fruit flies (Drosophila) have shown that the more highly developed species (mouse) is more sensitive than the less developed species (fruit fly). In fact, it appears that the mouse embryo is fifteen times more sensitive than the fruit fly embryo! Because data on humans are rare and dose can, at best, be estimated in these cases, it can be assumed that the human embryo is at least as radiosensitive as the mouse, if not more so.

OBSERVATIONS IN HUMANS.—Radiation effects on developing human embryos have been observed in the following situations: atomic bomb survivors, accidental exposures, occupational exposures and the diagnostic and therapeutic exposure of pregnant patients.

Congenital defects attributed to radiation in utero were described as early as 1930 by Murphy and Goldstein[8] in a report of microcephaly related to in utero exposure. A subsequent study of the children of 106 women who received irradiation for therapeutic purposes reported that 28 of 75 children had radiation-induced malformations. These children exhibited both central nervous system and skeletal defects including microcephalic idiocy, hydrocephaly, mental retardation without gross abnormalities, mongolism, spina bifida, double club-foot, limb deformities and blindness and other eye deformities. In all these cases radiation occurred during the first trimester.

Studies of children irradiated in utero at Hiroshima show that of 11 women who received a high dose of radiation,* 7 of their children were microcephalics and mentally retarded while children whose mothers were at a greater distance from the hypocenter and therefore were exposed to a lower dose did not exhibit an increased incidence of microcephaly. Of 30 children irradiated in utero at Nagasaki, there were 7 fetal deaths, 6 neonatal deaths and 4 mentally retarded children among the survivors.

In a study of the children of women irradiated with therapeutic doses during different stages of pregnancy, Dekaban[4] draws the following conclusions:

1. Over 250 R delivered to human embryos before 2–3 weeks of gestation may result in a large number of prenatal deaths but produce very few severe abnormalities in those children brought to term.
2. Irradiation of the human fetus between 4 and 11 weeks of gestation may lead to severe abnormalities of many organs, particularly the CNS and skeletal systems.
3. Irradiation during the eleventh and sixteenth weeks frequently produces mental retardation and microcephaly.
4. Although the fetus is more radioresistant in terms of lethality and abnormalities after the twentieth gestational week, irradiation during this time may result in functional defects.

Conclusions

In general it can be stated that the embryo and fetus are more sensitive to the effects of ionizing radiation than is the organism at any

---

*Although doses are difficult to estimate, these women showed clinical signs of exposure to high doses.

other period of life. In addition, there are variations in the radiation sensitivity during embryonic life. The first trimester, particularly the first 6 weeks of development, appears to be the most radiosensitive in terms of both lethality and induction of congenital abnormalities. The fetus becomes more resistant as development progresses through the second and third trimesters with higher doses necessary to produce damage.

Doses of 5 to 15 R that have been observed to be both fatal and to produce CNS abnormalities in mouse embryos during pre-implantation also may be damaging to human embryos during the first 2 weeks of development. However, because pregnancy in the human is generally not known or even suspected at this early stage and because the embryo may be resorbed by the body or aborted resulting in minimal, if any, indications of pregnancy, the implications to humans of these findings in mice are difficult to accurately assess.

The most radiosensitive time in the development of the human fetus for the induction of abnormalities is from the second through the sixth week, particularly the twenty-third through the thirty-seventh day of gestation. If irradiation occurs in this time interval, the greatest variety of abnormalities will be observed. As in the mouse, most radiation-induced congenital abnormalities in the human are related to the CNS. The most common abnormalities that have been observed in humans are microcephaly, mental retardation, sense organ damage and stunted growth. The third through the twentieth week of human gestation appears to be the most sensitive period for skeletal changes.

The sensitivity of the fetus during the first trimester is attributable to the large number of stem cells present during the early stages of development. The majority of abnormalities appear in the CNS and related sense organs due to the abundance and diffuseness of formative cells throughout the fetus and the exquisite radiosensitivity of these stem cells.

The second and third trimesters are more radioresistant than the first. Irradiation during the last two trimesters results in a lower incidence of abnormalities than irradiation during the first trimester. However, these latter stages of development may result in more subtle abnormalities and functional disorders (e.g., sterility) and late changes such as malignancies, particularly leukemia. In addition, children irradiated in utero with therapeutic doses during the last trimester may exhibit signs and symptoms of the bone marrow syndrome at birth.

The implications of fetal exposure in the medical uses of ionizing radiation will be discussed in following chapters.

REFERENCES

1. Bacq, Z. M.: *Fundamentals of Radiobiology* (2nd ed.; New York: Pergamon Press, 1961).
2. Behrens, C. S., *et al.*: *Atomic Medicine* (5th ed.; Baltimore: Williams & Wilkins, 1969).
3. Bond, V. P., *et al.*: *Mammalian Radiation Lethality: A Disturbance of Cellular Kinetics* (New York: Academic Press, 1965).
4. Dekaban, A. S.: Abnormalities in children exposed to x-radiation during various stages of gestation: Tentative timetable of radiation injury to the human fetus, J. Nucl. Med. 9:471, 1968.
5. Hempelmann, L. H., *et al.*: Acute radiation syndrome: Study of 9 cases and review of problem, Ann. Intern Med. 36:279, 1952.
6. Lushbaugh, C. C.: Reflections on Some Recent Progress in Human Radiobiology, in Augestein, L. G. (ed.): *Advances in Radiation Biology* (New York: Academic Press, 1969), pp. 277–314.
7. Murphy, D. P.: *Congenital Malformations* (Philadelphia: Lippincott, 1947).
8. Murphy, D. P., and Goldstein, L.: Micromelia in a child irradiated in utero, Surg. Gynecol. Obstet. 50:79, 1930.
9. Plummer, C.: Anomalies occurring in children exposed in utero to the atomic bomb at Hiroshima, Pediatrics 10:687, 1952.
10. Rugh, R., and Grupp, E.: Ionizing radiations and congenital anomalies in vertebrate embryos, Acta Embryol. Exp. 2:257, 1959.
11. Rugh, R.: Ionizing radiations; Their possible relation to the etiology of some congenital anomalies in human disorders, Milit. Med. 124:401, 1959.
12. Rugh, R.: Low levels of x-irradiation and the early mammalian embryo, Am. J. Roentgenol. Radium Ther. Nucl. Med. 87:559, 1962.
13. Rugh, R.: The impact of ionizing radiation on the embryo and fetus, Am. J. Roentgenol. Radium Ther. Nucl. Med. 89:182, 1963.
14. Rugh, R., and Wohlfromm, M.: Age of mother and previous breeding history and the incidence of x-ray induced congenital anomalies, Radiat. Res. 19:261, 1963.
15. Rugh, R.: Why radiobiology? Radiology 82:917, 1964.
16. Rugh, R.: X-ray induced teratogenesis in the mouse and its possible significance to man, Radiology 99:433, May 1971.
17. Rugh, R.: *From Conception to Birth; The Drama of Life's Beginnings* (New York: Harper & Row, 1971).
18. Russell, L. B., and Russell, W. L.: An analysis of the changing radiation response of the developing mouse embryo, J. Cell. Physiol. (Supplement 1) 43:103, 1954.
19. Russell, L. B., and Montgomery, C. S.: Radiation sensitivity differences with cell-division cycles during mouse cleavage, Int. J. Radiat. Biol. 10:151, 1966.
20. Thoma, Jr., G. E., and Wald, N.: Acute Radiation Syndrome in Man, in *Fundamentals of Radiological Health*, Training Manual of the National Center for Radiological Health, DHEW Training Publication No. 3n.
21. Wald, N., *et al.*: Hematologic manifestations of radiation exposure in man, Prog. Hematology 3:1, 1962.
22. Yamazaki, J. N., *et al.*: Outcome of pregnancy in women exposed to the atomic bomb in Nagasaki, J. Dis. Child. 87:448, 1954.

7/Late Effects of Radiation

The previous chapters were concerned with the immediately lethal effects induced in the adult and fetus by an acute high dose of radiation (greater than 100 rads). Although knowledge of these lethal effects is necessary, incidents resulting in exposure of a magnitude that causes death are relatively few. Of equal interest are the effects observed in individuals years later who survive these acute doses. Because these effects (the damage produced by irradiation) are insidious and are manifested after long periods of time, they are termed "late effects" of radiation. Unlike the immediate effects of an acute high dose, late effects remain dormant for many years and, in fact, may not be seen in the individual but in succeeding generations. This delay (or latent period) in the manifestation of late effects renders this area of radiobiology more difficult to observe and evaluate than acute effects in both experimental and actual situations.

Of particular interest and concern is the fact that late effects may be induced by low doses and chronic low doses of radiation (i.e., low doses given over a long period of time), such as those received by patients in diagnostic radiology and nuclear medicine or by occupationally exposed persons.

The late effects of radiation discussed in this chapter include somatic and genetic effects.

Somatic Effects

Somatic effects are defined as those which affect the health of the individual; late somatic effects induced by radiation include carcinogenesis and nonspecific life shortening.

Carcinogenesis

Radiation has long been known to be a carcinogenic (cancer-inducing) agent. The first reported case of radiation-induced carcinoma was in 1902 on the hand of a technician. Within 15 years of the discovery of x-rays, 100 cases of skin cancer caused by radiation were reported in occupationally exposed personnel, both radiologists and technicians. Since that time, radiation has been implicated as the cause of other types of malignancies.

METHODS OF STUDYING RADIATION-INDUCED MALIGNANCIES.—Radiation has been implicated as an etiologic (causative) factor for cancer primarily through laboratory animal studies and statistical studies of human populations exposed to radiation. Incidence rates for cancer induced by radiation are determined by comparing the expected incidence of cancer in the control group (i.e., the general population) to the incidence in the experimental group (i.e., the irradiated population) and calculating risk factors for the irradiated population. A number of problems arise from the statistical nature of these studies, termed epidemiologic studies (studies of disease incidence in human populations). Three of these problems are as follows.

First, the observed incidence of the disease in the experimental (in this case irradiated) population is compared to the expected incidence in the general population. This comparison is certainly valid when the experimental population consists of nondiseased individuals (e.g., studies on radiologists). However, an experimental population consisting of diseased individuals (e.g., individuals treated with radiation for ankylosing spondylitis) may not constitute a valid comparison.

Second, the actual number of persons with cancer in the general population appears to be large but, when compared to the total number of persons in the general population, individuals with cancer constitute a small segment. Because of these small numbers, an increase in the number of cancer cases, although appearing to be large, may not be statistically significant.

Third, studies of this type cannot, by their nature, exclude other factors that may play a part in causing the observed increase in a specific disease.

With these thoughts and precautions in mind, the first section of this chapter presents the statistical evidence implicating radiation as an etiologic factor in cancer. Because the incidence of radiation-induced malignancies in humans can be documented, this section will briefly discuss these experiences.

RADIATION-INDUCED MALIGNANCIES IN HUMANS.—The following are sources of data on the incidence of radiation-induced cancer:
1. Occupational exposure.
2. Atomic bomb survivors.
3. Medical exposure.
4. Fallout accidents in the Pacific Testing Grounds.

Malignancies in which radiation has been implicated as a cause are leukemia, skin carcinoma, osteosarcoma (cancer of the bone) and lung and thyroid carcinoma. Probably the two malignancies most often associated with radiation are leukemia and skin carcinoma.

LEUKEMIA.—The role of radiation as a cause of leukemia was first noted in 1911 in a report of eleven cases in which radiation was implicated as being leukemogenic (leukemia causing) in occupationally exposed individuals. Since that time, radiation has been definitely linked to leukemia by studies of adult populations such as the atomic bomb survivors, patients treated with radiation for ankylosing spondylitis and radiologists. In addition, some studies tentatively have linked in utero exposure to diagnostic doses of radiation to the later incidence of leukemia.

Studies of the atomic bomb survivors in both Hiroshima and Nagasaki show a statistically significant increase in leukemia incidence in the exposed population as compared to the nonexposed population. In the period 1950–1956, 117 new cases of leukemia were reported in the Japanese survivors; approximately 64 of these can be attributed to radiation. In Hiroshima there was an observed frequency of 61 leukemia deaths compared to an expected incidence of 12—a five-fold increase. In Nagasaki there were 20 deaths attributable to leukemia compared to an expected rate of 7—a three-fold increase. Even today an increased risk of leukemia exists in this population compared to the nonirradiated population.

An increased leukemia incidence also has been observed in the population exposed to low doses (20–50 rads) at Hiroshima but not at Nagasaki. This observed difference in leukemia incidence in the two populations may be due to the different types of radiation to which they were exposed; at Nagasaki, approximately 90% of the dose was due to x-rays while at Hiroshima about half the dose was from x-rays and half from neutrons.*

Patients with ankylosing spondylitis treated with radiation in Great Britain comprise the second major population implicating radiation as being leukemogenic. From 1935 to 1944, approximately 15,000 patients were irradiated with both acute and fractionated doses ranging from 100 to 2000 R, often to the spine and pelvis. Two-year follow-up of the patients revealed seven cases of leukemia compared to an expected incidence of one. An increase in leukemia incidence of over 100% was observed among patients who received doses greater than 2000 R. In these studies the population observed was initially ill and consisted primarily of males† therefore resulting in a possible overestimate of risk when compared to the whole population.

Another group that implicates radiation as a causal factor in leukemia comprises the early radiologists. A study of 425 radiologists in

*Neutrons have a higher RBE than x-rays, i.e., neutrons are biologically more damaging than x-rays.

†Males appear to be more susceptible than females.

the United States who died between 1948 and 1961 revealed an incidence of 12 cases of leukemia compared to an expected incidence of 4 cases; all were of the types known to be increased by radiation. Doses are difficult to estimate in these individuals but range from 100 rem* up to 2000 rem accumulated over each lifetime of practice. The incidence in American radiologists was lower than in the Japanese population, which may be accounted for by differences in dose rate (high-dose rate: Japanese; low-dose rate: radiologists), the volume of the body exposed (total body or near total body: Japanese; partial body: radiologists) and acute (Japanese) versus chronic (radiologists) exposure.

A study of British radiologists entering practice after 1921 revealed no excess in leukemia incidence thereby refuting the findings in American radiologists.

Whether this increased leukemia incidence in American radiologists has returned to normal at this time remains controversial; some studies report normal incidence while others report an increased incidence compared to other medical specialists. One reason for this observed increase may be the inclusion of radiologists in the sample population who practiced before 1955 when the current permissible doses were established. It is now accepted that the incidence of leukemia in radiologists is not higher than in other medical specialists or in the general population.

Radiation during fetal life has long been implicated as a cause of cancer. Doses of 1 to 5 rads during the prenatal period *appear* to result in an increase in *all* types of cancer, particularly leukemia. Because these diseases are rare, the numbers in the studies are small and may not be significant. However, a study with a large sample population (i.e., 750,000 children) and more recent in-depth studies of this population of children indicate that the leukemogenic effect of radiation may be real, resulting in as much as an eightfold increase of leukemia in children irradiated in utero in contrast with children not irradiated in utero. This latter study is of particular importance because the irradiated population was matched with the nonirradiated population in terms of other diseases, removing the possibility that other disease factors were contributing to the increased incidence.

In all samples the population irradiated in utero had a higher leukemia incidence than the matched control population. If these numbers are real, the risk of leukemia as a result of prenatal exposure is

---

*A rem is a unit of dose equivalent used in radiation protection. It is defined as the dose in rads x quality factor (QF) times modifying factors (MF). QF allows for the different biologic effects of different types of radiation and is similar to RBE; MF applies primarily to internally deposited radionuclides and accounts for variations in the mode and site of deposition.

much greater than postnatal exposure. However, these individuals may not be representative of the whole population, and other factors as yet undefined may be contributing to the increased leukemia incidence. Studies of the children of exposed Japanese women who were pregnant at the time of the bombing have observed no increase in leukemia at this time, further substantiating this possibility. Although this topic remains controversial, evidence is accumulating that in utero radiation exposure, although not the sole factor, may be a contributing factor to the induction of leukemia.

Because radiation-induced leukemia is of more concern and is more controversial in contemporary life than other radiation-induced cancers, some pertinent questions and conclusions will be summarized at this point.

The latent period for leukemia induction following radiation exposure is approximately 1 to 2 years with a peak incidence at 5 to 7 years, declining after 15 years to the expected incidence. An interesting point is that males appear to have a higher susceptibility to radiation-induced leukemia than do females.

Specific cell types of leukemia are related to radiation exposure; adult exposure results in an increase of the acute and chronic myeloid types, while childhood exposure results in an increase of acute lymphocytic leukemia. There also appears to be a dose-rate effect based on the comparison of leukemia incidence in the American radiologists and the Japanese survivors exposed to low and high dose-rates, respectively.

Of particular importance is the question of dose—at what dose of radiation is there a high probability of leukemia occurring? Based on the above studies, an acute dose of 750 rads to adults, particularly if given to a large volume of red bone marrow, greatly increases the incidence of the disease. An even more important and pertinent medical question concerns low doses and the existence of a threshold dose—questions that are much more controversial and difficult to answer. The threshold concept (i.e., below a certain dose there is no increased incidence of leukemia) is supported by the decreased incidence of leukemia in radiologists entering practice after 1950 when maximum permissible doses (MPD's) were lowered to their present limits. Based on this evidence, doses of a few rads appear to have a small risk of inducing leukemia in adults. However, studies of the Hiroshima population exposed to low doses suggest the possibility that radiation-induced leukemia may be a nonthreshold response, i.e., any dose, no matter how small, may produce an increased risk of the disease. Studies reporting an increased leukemia incidence in mice exposed to 10 R further support the nonthreshold dose concept for leukemia induction by radiation (Fig. 7-1).

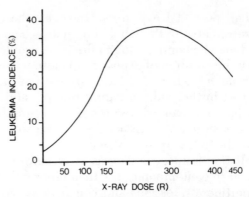

Fig. 7-1.—Incidence of leukemia in male mice exposed to total body irradiation. Note the decrease in leukemia incidence after exposures greater than 300 R which is due to the increasing incidence of fatalities at this and higher doses. (From Upton, A. C.: Cancer Res. 21:717, 1961.)

Although no conclusive answer is presently available, more evidence is accumulating that suggests that leukemia induction by radiation may be a nonthreshold linear response, i.e., equal increases in dose cause a corresponding equal increase in incidence of the disease. However, one important point that must be remembered is that although radiation-induced leukemia may be a nonthreshold response, the linearity of the response results in a much smaller risk at low doses than at high doses; in fact, the risk at low doses may be so small as to be unidentifiable from other contributing factors.

SKIN CARCINOMA.—Several radiation reactions were reported on the hands of early radiologists and technicians soon after Röntgen's discovery, with the first case of skin cancer reported in 1902 on the hand of a technician. Because their x-ray machines were crude, radiologists placed their hands in the beam to check the efficiency, resulting in high exposure to the hands with a large number of individuals developing skin carcinoma years later. In fact, 100 cases of skin cancer had been reported in radiologists and technicians 15 years after the discovery of x-rays. Another source of information regarding radiation-induced skin cancer is individuals treated with radiation for acne and ringworm; many developed skin cancer years later.

Although skin carcinoma caused by radiation cannot be differentiated from skin cancer caused by other agents, enough evidence exists to implicate radiation as a definite cause. What must be kept in mind concerning both these groups is that they were exposed to unfiltered low Kv x-rays that produce extremely high doses in the superficial layers of the skin. Because of these incidents in pioneer workers that resulted in safety precautions on x-ray units and a reduction in MPD, radiation-induced skin cancer has disappeared in occupationally exposed persons.

OSTEOSARCOMA (BONE CANCER).—The story of the watch dial painters in the early part of this century (1915–1930) who pointed their brushes with their lips is well known. Radium, used to paint the clock faces, was ingested by these workers in the above manner and, being a bone-seeker, accumulated in the skeleton. Although exposure to the radioactive material was brief, the ingested radium was not excreted resulting in continuous exposure to the bone. As a result, approximately 40 of a total of several hundred persons developed osteosarcoma—a much greater than expected incidence.

LUNG CARCINOMA.—Pitchblende miners in Germany more than 500 years ago were known to have a high incidence of a fatal disease referred to as "mountain sickness." In 1924 investigation of this disease revealed it to be carcinoma of the lung. The cause of the disease was occupational and partially attributable to the fact that the air in the pitchblende mines was rich in radon.

An increased incidence of lung cancer was reported among uranium miners in the western United States. A study of miners from 1950 to 1967 revealed 62 cases of lung cancer, 6 times the expected number. This number was also increased over the number of lung cancers reported in other types of miners. These tumors are believed to be caused by the inhalation and subsequent deposition in the lungs of dust containing radioactive material.

THYROID CARCINOMA.—The practice of irradiating the thymus in infants to reduce "thymic enlargement" gained popularity in the early decades of this century and declined after the 1930's. Doses to these infants ranged from 120 to 6000 R, resulting in a hundredfold increase in the incidence of thyroid cancer. Although there is not sufficient data on children with the same disease who were not irradiated to allow a valid comparison of the effects induced by radiation and those caused by the disease itself, all studies of this group reveal a significantly higher incidence of thyroid cancer in the irradiated than the nonirradiated population.

Another source of information concerning thyroid cancer attributable to radiation is the children of the Marshall Islanders who were accidentally exposed to fallout radiation from a nuclear test device when a sudden wind shift occurred at the time of testing. In later years these irradiated children revealed an increased incidence of all types of thyroid disease, including benign and malignant tumors. Follow-up studies of the survivors at Hiroshima and Nagasaki also have observed an increased incidence of thyroid cancer in those individuals who were children at the time of the bombing. Although doses are difficult to determine in the above situations, estimates reveal that a dose of 100 rads or less had been received by some individuals who later developed thyroid cancer. The latent period for this radiation-induced cancer appears to be 10 to 20 years.

OTHER MALIGNANCIES.—Although the above cancers are those most often attributed to radiation, it also has been implicated as an etiologic factor for other cancers such as breast carcinoma, salivary gland cancer and some sarcomas.

CONCLUSIONS.—The fact that radiation is an etiologic factor in cancer cannot be disputed. The major controversy concerns dose information; in most instances, only estimates are available. Doses as low as 25 rads have been implicated as causing malignancies in adults whereas in utero exposures are lower. The fact that infants and children are generally more radiosensitive may imply that lower doses will cause more cancers in the young than in adults; an even lower dose may cause cancer in the radiosensitive fetus.

Latent periods vary with the type of cancer studied and range from 1 to 30 years. The latent period for the manifestation of a specific radiation-induced cancer is dose-dependent as are all radiation effects, i.e., with increasing dose, the latent period decreases and the disease appears earlier. Because of the unanswerable doubts regarding dose, latent periods also are difficult to estimate.

The question of a threshold versus a nonthreshold linear response remains controversial, although more evidence is accumulating that supports the nonthreshold concept for all radiation-induced cancers. If the threshold concept is proved, different malignancies will probably exhibit different threshold doses. However, if radiation-induced cancer is a nonthreshold response, the risk at low doses may be so small as to be unidentifiable from other environmental factors that also may be responsible for an increased cancer incidence.

Today, in general, the incidence of radiation-induced cancer appears to be decreasing—a trend that may be attributable to two factors:
1. The knowledge of these dangerous effects, which has led to strict regulations regarding usage and a lowering of MPD's to occupationally exposed personnel and the general population.
2. The disappearance of the misuse of radiation as a panacea for treating illnesses of all types.

How Does Radiation Induce Cancer?

Although radiation has been shown to increase the incidence of cancer, the reasons for its doing so remain speculative. Structural changes in chromosomes resulting in mutations are a known effect of radiation. Although it is certainly possible that a mutated cell may be the responsible agent, the probability of a mutation in one or a few cells resulting in a malignancy is small.

Other theories proposed concerning the induction of cancer by radiation include:
1. A mutation in a somatic cell leading to uncontrolled cell growth.
2. Acceleration of aging with a concomitant high incidence of cancer.
3. Altered cellular environment.
4. Abnormal cell division following repeated attempts at repair of radiation injury.

It will more likely be found that a combination of all of the above play a role in inducing cancer—not just one. In addition, other environmental and health factors that increase the risk of cancer have been identified; more of these undoubtedly will be identified in the future. The present evidence implies that radiation is one of many insults, not the sole factor, in the etiology of cancer, particularly in view of current low exposures to all individuals. The mechanism by which radiation induces cancer remains elusive and controversial.

Nonspecific Life Shortening

Studies in small animals have shown that animals chronically exposed to low doses of radiation die younger than animals never exposed to radiation (Fig. 7-2). Examination of these animals at death revealed a decrease in the number of parenchymal cells and blood vessels and an increase in connective tissue in organs—indications of an aging process. This phenomenon is often referred to as radiation-induced aging because it appears to be an acceleration of the aging process.

Nonspecific life shortening due to radiation also has been reported in humans, although it appears to be accounted for by an increase in malignancies, particularly in the low-dose range. This is a

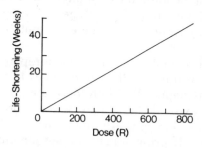

Fig. 7-2.—Lifespan shortening in mice as a function of radiation dose; note direct relationship between life-shortening and dose. (From Rotblat, J., and Lindop, P.: Proc. R. Soc. Lond. [Biol.] 154:332, 1961. Courtesy of the Royal Society.)

controversial area, because some studies indicate a shortening of lifespan while others indicate the opposite, a lengthening of lifespan.

Genetic Effects

Although somatic late effects are unfortunate for the afflicted individuals, of equal if not greater consequence is the impact of radiation on future generations. In fact, for many years the hazards and risks to future generations were believed to be of much greater consequence than those to the individual—a belief that is being questioned today.

Genetics and the function of DNA as the storehouse of all genetic information were discussed in Chapter 1. It was also pointed out that genes and the sequence of bases on the DNA chain are stable, i.e., during both mitosis and meiosis, DNA reproduces a duplicate of itself which is transmitted to the daughter cells thus maintaining the integrity of genetic information in the cell. This process is extremely important in germ cells which transmit the information to future generations.

Occasionally, for some unknown reason genes and DNA spontaneously change, altering either the structure or the amount of DNA in the cell. These naturally occurring changes, termed *spontaneous mutations*, are permanent and heritable, i.e., they may be passed from cell to cell and possibly from generation to generation.

Generally thought to be detrimental, the effect of mutations on the individual are dependent on the gene in which the change has occurred. Some genes are vital to life, therefore a mutation in these genes is severely detrimental, possibly resulting in the death of the individual before adulthood. Mutations in less vital genes are of less consequence to the life of the individual. Examples of spontaneous mutations in the human population are mongoloids and hydrocephalics.

A certain number of spontaneous mutations arise in each generation; this is termed the *mutation frequency*. The frequency of spontaneous mutations in a generation can be altered by a number of factors including viruses, chemicals and radiation. These agents, termed *mutagens*, are responsible for increasing the spontaneous mutation rate.

Radiation certainly can produce mutations through unrepaired structural breaks in chromosomes or through discrete changes in the order of bases on the DNA chain (Chapter 2). When these mutations occur in germ cells, the possibility exists that they may be transmitted to future generations. The consequences of these mutations may be manifested in the first generation of offspring or may not be seen until the second, third or even later generations. Chapter 6 discussed the effects of in utero irradiation on the fetus; this section will discuss the second way in which radiation affects the fetus and therefore future generations through the transmission of a radiation-induced mutation in the ovum or sperm. This category comprises the genetic effects of radiation.

Methods of Studying Radiation-Induced Mutations

The effect of radiation on mutation frequency has been studied extensively in laboratory animals—primarily the fruit fly and mouse. As in studying other effects of radiation, it is not possible to subject humans to the controlled conditions necessary for determining these effects. In the study of genetic effects, breeding is controlled, i.e., only certain males are permitted to mate with specific females because of their characteristics. The offspring are then studied for the presence or absence of these characteristics. A second important factor in studying genetic effects is the lifespan of the species. Because genetic effects are not necessarily exhibited in the first generation of offspring, many generations must be observed, necessitating a short lifespan. Most laboratory animals used for genetic studies fit this criterion. In addition, the laboratory animals used give birth to multiple offspring per mating—an unusual situation in humans. This larger number of offspring increases the probability of observing genetic effects.

Study of Radiation Genetics in Animals

The classic study of the effect of radiation on mutation frequency was done in 1927 by Herman J. Müller in a comprehensive study of fruit flies.[29] By exposing males and females to various doses of radiation, observing lethality and the change in appearance of the offspring of irradiated parents and comparing these results with unirradiated controls, Müller drew the following conclusions:

1. Radiation does not produce any new or unique mutations—it simply increases the number of mutations that spontaneously arise in each generation.
2. Between doses of 25 and 400 R, the frequency of mutations was linear with dose, i.e., equal dose increments produced an equal increase in the number of mutations.
3. Radiation-induced mutations are recessive,* i.e., they may not be exhibited for many generations.

---

*For recessive mutations to be expressed in an individual, the same mutated gene must be received from both parents. Although this is an unlikely event, constant exposure of a population to a mutagen such as ionizing radiation increases the probability that recessive mutations will be expressed in future generations.

From Müller's and other studies, generalizations have been drawn concerning radiation effects on mutation frequency. One of these is that, theoretically, there is no dose too small to produce a mutation. As a result, the production of mutations is said to be a nonthreshold response. This was postulated because of the linear relationship exhibited between dose and mutation frequency in Müller's study. Although the lowest dose used by Müller was 25 R, other studies have observed the same linear response using doses as low as 5 R. However, this is a controversial area; some individuals maintain that a threshold dose exists below which the mutation frequency is not increased. To be on the safe side until conclusive evidence is presented to support the threshold concept, genetic effects must still be assumed to be a nonthreshold linear response with low doses producing as much as one-tenth the genetic damage as high doses (Fig. 7-3).

In all experimental studies, radiation has not produced any new mutations but has simply increased the number that arise spontaneously. In addition, those mutations produced by radiation are recessive and may not be manifested for many generations.

Another observation was that high LET radiations are much more efficient for producing mutations. This is to be expected because of the physical characteristics of the radiation and the fact that chromosome breaks do indeed exhibit a dependence on LET (Chapter 2).

In addition, dose-rate dependence has also been noted; low-dose rates produce less genetic damage than high-dose rates.

Doubling Dose

The unit of measurement for the determination of radiation effect

Fig. 7-3.—The percentage of recessive lethal mutations in the fruit fly as a function of radiation dose. (From Hall, E. J.: *Radiobiology for the Radiologist* [New York: Harper & Row, 1973], p. 222.)

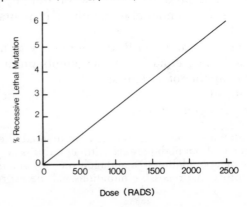

on mutation frequency is the *doubling dose*—defined as that dose of radiation which ultimately doubles the number of spontaneous mutations. For example, if 5% of the offspring in each generation are observed to be mutations, the doubling dose would eventually produce 10% mutations. Because radiation produces no new mutations, only an increase in the already existing number of mutations is observed.

The doubling dose is estimated to be between 30 and 80 R and is generally given as 50 R. A dose range is given because a difference in mutation frequency is observed with chronic and acute exposures. The doubling dose for chronic exposure is 80 R, for acute it is 30 R. In practice, what doubling dose means is if each individual of child-bearing potential in the whole population were exposed to a gonadal dose of 50 R for many generations, eventually there would be twice the number of mutations in the offspring of these individuals.

Genetic Effects in Humans

Because the study of the effects of radiation on mutation rate in humans is not feasible, most information concerning these effects has been extrapolated from animal studies. In 1955 Neel made the following comment concerning this attempted extrapolation: "Many of the conclusions derived from experiments with Drosophila (fruit fly) and mice ... cannot be interpolated because man is neither an overgrown fruit fly nor an oversized mouse."[31]

One situation in humans that has allowed the observation of radiation-induced genetic effects is the study of children conceived after one or both parents had been exposed to radiation at Hiroshima or Nagasaki. Approximately 71,000 pregnancies were entered in a study in postwar Hiroshima and Nagasaki between 1948 and 1953; observation of the children revealed no effect on the six genetic factors studied.

Therefore, although laboratory experiments have revealed that radiation is indeed a mutagenic agent, the effect was not observed in the first generation offspring of the Hiroshima and Nagasaki survivors. This does not indicate that mutations have not occurred—they simply may not be exhibited in the first generation offspring. There is no assurance that mutations will not be observed in the second or third generations, particularly since radiation-induced mutations are recessive. In addition, the number of people exposed was small by genetic standards.

Conclusion

Many of the late effects discussed in this chapter are no longer evident in the population and are therefore of historical interest.

Much of the information on the late somatic effects of radiation is derived from observations many years ago, when exposures by today's standards were high and before radiation protection was a glimmer in anyone's eye. In fact, it is primarily due to the early pioneers in the field that we owe our current knowledge of the late effects of this potentially dangerous tool. This knowledge is due not to the carelessness of these individuals but to their ignorance of the powerful tool in their hands.

With the advent of maximum permissible doses (Chapter 8) and the continual lowering of these limits to their present levels (5 rem/year), most late somatic effects are considered to be of historic interest, provided, of course, that all safety rules and regulations regarding exposure to ionizing radiation are followed. Genetic effects, for many years considered to be the most potentially hazardous late effect, may be of less risk and consequence than the risk of somatic effects.

This chapter, some of which is historic in nature, is primarily intended to reaffirm to the reader that ionizing radiation is dangerous, even at low doses, when safety rules and regulations are neglected. The implications of late effects both to the individual (whether general population, patient or occupationally exposed person) and to future generations, makes this an extremely important area of which far too little knowledge exists, particularly in view of the increased use of ionizing radiation for medical purposes.

REFERENCES

1. Albert, R. W., *et al.*: Follow-up study of patients treated by x-ray epilation for tinea capitis, Arch. Environ. Health 17:899, 1968.
2. Bross, I. D., and Natarajan, N.: Leukemia from low level radiation: Identification of susceptible children, N. Engl. J. Med. 287:107, 1972.
3. Conard, R. A., *et al.*: Thyroid nodules as a late sequela of radioactive fallout, in a Marshall Island population exposed in 1954, N. Engl. J. Med. 274:1391, 1966.
4. Conard, R. A., *et al.*: Thyroid neoplasia as late effect of exposure to radioactive iodine in fallout, J.A.M.A. 214:316, 1970.
5. Conard, R. A., and Hicking, A.: Medical findings in Marshallese people exposed to fallout radiation: Results from a ten-year study, J.A.M.A. 192:457, 1965.
6. *Court-Brown, W. M., and Doll, R.: Expectation of life and mortality from cancer among British radiologists, Br. Med. J. 2(5090):181, 1958.
7. Court-Brown, W. M., *et al.*: The incidence of leukaemia after the exposure to diagnostic radiation in utero, Br. Med. J. 2(5212):1599, 1960.
8. Court-Brown, W. M., and Doll, R.: Mortality from cancer and other causes after radiotherapy for ankylosing spondylitis, Br. Med. J. 2(5474):1327, 1965.

---

*Court-Brown, W. M., is occasionally indexed as Brown, W. M.

9. Curtis, H. J.: Radiation-Induced Aging in Mice, *Third Australasian Conference on Radiobiology, University of Sydney, 1958* (London: Butterworth, 1961), pp. 114—22.
10. Dublin, L. I., and Spiegelman, M.: Mortality of medical specialists, 1938–1942, J.A.M.A. 137:1519, 1948.
11. *Effects on Populations of Exposures to Low Levels of Ionizing Radiations,* Report of the Advisory Commission on the Biological Effects of Ionizing Radiations, Division of Medical Sciences, National Academy of Sciences/National Research Council, November 1972.
12. Evans, R. D., *et al.*: Radiogenic tumors in the radium and mesothorium cases studied at MIT, in May, C. W., *et al.* (eds.): *Delayed Effects of Bone-Seeking Radionuclides* (Salt Lake City: University of Utah Press, 1969), p. 157.
13. Ford, D. D., *et al.*: Fetal exposure to diagnostic x-rays and leukemia and other malignant diseases in childhood, J. Natl. Cancer Inst. 22:1093, 1959.
14. Griem, M. L., *et al.*: Analysis of the morbidity and mortality of children irradiated in fetal life, Radiology 88:347, 1967.
15. Härting, F. H., and Hesse, W.: Der Lungekrebs, die Berkrankheit in den Schneeeberger Gruben, Vrljschr. Gerichtl. Med., Berlin, 1879, n. F., 30:296; 31:102, 313.
16. Hall, E. J.: *Radiobiology for the Radiologist* (New York: Harper & Row, 1973).
17. Hempelmann, L. H.: Risk of thyroid neoplasms after irradiation in childhood; Studies of populations exposed to radiation in childhood show a dose response over a wide dose range, Science 160:159, 1968.
18. Ju, D. M.: Salivary gland tumors occurring after irradiation of the head and neck area, Am. J. Surg. 116:518, 1968.
19. Krall, J. F.: Estimation of spontaneous and radiation induced mutation rates in man, Eugenics Q. 3:201, 1956.
20. McKenzie, I.: Breast cancer following multiple fluoroscopies, Br. J. Cancer 19:1, 1965.
21. MacMahon, B., and Newill, V. A.: Birth characteristics of children dying of malignant neoplasms, J. Natl. Cancer Inst. 28:231, 1962.
22. MacMahon, B.: Prenatal x-ray exposure and childhood cancer, J. Natl. Cancer Inst. 28:1173, 1962.
23. March, H. C.: Leukemia in radiologists, Radiology 43:275, 1944.
24. March, H. C.: Leukemia in radiologists in a 20-year period, Am. J. Med. Sci. 220:282, 1950.
25. Martland, H. S.: Occurrence of malignancy in radioactive persons; general review of data gathered in study of radium dial painters, with special reference to occurrence of osteogenic sarcoma and interrelationship of certain blood diseases, Am. J. Cancer 15:2435, 1931.
26. Miller, R. W.: Delayed radiation effects in atomic-bomb survivors; Major observations by the Atomic Bomb Casualty Commission are evaluated, Science 166:569, 1969.
27. Mole, R. H.: Radiation effects in man: Current views and prospects, Health Phys. 20:485, 1971.
28. Morgan, K. Z.: Biological effects of ionizing radiation; Lecture given at course "Environmental Analysis and Environmental Monitoring for Nuclear Power Generation," University of California, Berkeley, September 9—13, 1974.
29. Müller, H. J.: Artificial transmutation of the gene, Science 66:84, 1927.

30. Neel, J. V.: On some pitfalls in developing an adequate genetic hypothesis, Am. J. Hum. Genet. 7(1):1,1955.

31. Neel, J. V.: *Changing Perspectives on the Genetic Effects of Radiation* (Springfield, Illinois: Charles C Thomas, 1963).

32. Pack, G. T., and Davis, J.: Radiation cancer of the skin, Radiology 84:436, 1965.

33. Petersen, O.: Radiation cancer: Report of 21 cases, Acta Radiol. (Stockh.) 42 (3):221, 1954.

34. Rotblat, J., and Lindop, P.: Long-term effects of a single whole body exposure of mice to ionizing radiation, II. Causes of death, Proc. R. Soc. Lond. [Biol.] 154:350, 1961.

35. Russell, W. L.: Genetic hazards of radiation, Proc. Am. Phil. Soc. 107:11, 1963.

36. Russell, W. L.: Studies in mammalian radiation genetics, Nucleonics 23:53, 1965.

37. Saccomanno, G., et al.: Lung cancer of uranium miners on the Colorado plateau, Health Phys. 10:1195, 1964.

38. Seltser, R., and Sartwell, P. E.: Influence of occupational exposure to radiation on mortality of American radiologists and other medical specialists, Am. J. Epidemiol. 81:2, 1965.

39. Simpson, C. L., and Hempelmann, L. H.: The association of tumors and roentgen-ray treatment of the thorax in infancy, Cancer 10(1):42, 1957.

40. Socolow, E. L., et al.: Thyroid carcinoma in man after exposure to ionizing radiation; A summary of the findings in Hiroshima and Nagasaki, N. Engl. J. Med. 268:406, 1963.

41. Spencer, W. P., and Stern, C.: Experiments to test the validity of the linear R dose mutation frequency relationship in Drosophila at low dosage, Genetics 33:43, 1948.

42. Sternglass, E. J.: Evidence for Low Level Radiation Effects on the Human Embryo and Fetus, *Proceedings of the Ninth Hannford Biology Symposium*, Atomic Energy Commission Symposium Series 17 (1969), pp. 651-60.

43. Stewart, A.: An epidemiologist takes a look at radiation risks, *USDHEW Publication (FDA) 73-8024*, January 1973.

44. Stewart, A., et al.: A survey of childhood malignancies, Br. Med. J. 1(5086): 1495, 1958.

45. Sutow, W. W., and Conard, R. A.: Effects of Fallout Radiation on Marshallese Children, *Proceedings of the Ninth Hannford Biology Symposium*, Atomic Energy Commission Symposium Series 17 (1969), pp. 661–74.

46. Toyooka, E. T., et al.: Neoplasms in children treated with x-rays for thymic enlargement, II. Tumor incidence as a function of radiation factors, J. Natl. Cancer Inst. 31:1357, 1963.

47. U.S. Congress, Joint Committee on Atomic Energy, Subcommittee on Research Development in Radiation: *Radiation Exposure of Uranium Miners* (Washington, D.C.: U.S. Government Printing Office, 1968).

48. Upton, A. C.: *Radiation carcinogenesis*, in *Methods in Cancer Research*, Vol. 4 (New York: Academic Press, 1968).

49. Wanebo, C. K., et al.: Breast cancer after exposure to the atomic bombings of Hiroshima and Nagasaki, N. Engl. J. Med. 279:667, 1968.

50. Warren, S.: Longevity and causes of death from irradiation in physicians, J.A.M.A. 162(5):464, 1956.

51. Warren, S., and Lombard, O. M.: New data on the effects of ionizing radiation on radiologists, Arch. Environ. Health (Chicago) 13:415, 1966.
52. Warren, S.: Radiation Carcinogenesis. Bull. N.Y. Acad. Med. 46:131, 1970.
53. Wolff, S.: Radiation genetics, Annu. Rev. Genet. 1:221, 1967.

8/Clinical Radiobiology I: Diagnostic Radiology

The general population is exposed to ionizing radiation from two sources—natural and human-made. Natural sources of radiation include exposure from the earth's crust, outer space, building materials and naturally occurring radioactive materials in the body. Human-made radiation sources are medical (including diagnostic radiography, nuclear medicine and radiation therapy) and dental exposures, fallout from nuclear weapons, nuclear power industries and occupational exposures. The contribution from both natural and human-made sources to total body and gonadal exposure of the general population is given in Table 8-1. Although natural radiation sources make the largest contribution to exposure of the general population, exposure from human-made sources is following closely. The largest single contributor of radiation to the general population from human-made sources is medical/dental exposures.

In 1970, 130 million people (65% of the general population of the United States) were exposed to medical and dental x-rays. In comparison to 1964, this figure represents an increase of 20% in the number of persons exposed to medical and dental radiation, versus a 7% population increase. Table 8-2 presents the types of medical/dental procedures and the approximate number of persons exposed to these procedures in 1970. As seen from the table, medical diagnostic radiology (i.e., radiographic and fluoroscopic procedures) is responsible for the majority of radiation exposures.

In view of these figures, it is interesting to reflect on an editorial in the *Pall Mall Gazette* (London, 1896) concerning x-rays and their use:

We are sick of the röntgen ray ... you can see other people's bones with the naked eye, and also see through eight inches of solid wood. On the revolting indecency of this there is no need to dwell. But what we seriously put before the attention of the Government ... that it will call for legislative restriction of the severest kind. Perhaps the best thing would be for all civilized nations to combine to burn all works on the röntgen rays, to execute all the discoverers, and to corner all the tungstate in the world and whelm it in the middle of the ocean.

It is indeed fortunate that this feeling did not gain widespread acceptance, because the use of radiation in the diagnosis of disease

TABLE 8-1.—SOURCES OF GENETICALLY SIGNIFICANT
RADIATION; ESTIMATED AVERAGE AMOUNTS*

SOURCE	WHOLE BODY (MREM/YEAR)	GENETICALLY SIGNIFICANT EXPOSURE (MREM/YEAR)
Natural Radiation		
Cosmic radiation	44	
Radionuclides in body	18	
External γ-radiation	40	
Total	102	90
Man-made radiation		
Medical/dental	73	30–60
Fallout	4	
Occupational exposure	0.8	
Nuclear power (1970)	0.003	
Nuclear power (2000)	<1	
Total	Approximately 80	

*From *Effects on Populations of Exposure to Low Levels of Ionizing Radiation,* Report of the Advisory Committee on the Biological Effects of Ionizing Radiation, Division of Medical Sciences, National Academy of Sciences/National Research Council (Washington, D.C.: U.S. Government Printing Office 1972). Courtesy of the National Academy of Sciences.

has become an invaluable and necessary tool in the armamentarium of every physician, regardless of specialty. At this time there is no reason to assume that the use of radiation for diagnostic purposes will decrease; in fact, continued growth is to be expected, with a greater number of persons exposed to a greater variety of procedures.

Although the benefits derived from diagnostic procedures cannot be disputed, the risks involved to both patients and personnel and the principles of radiation hygiene must be considered by all individuals administering radiation, both physicians and technologists.

From all available evidence, immediate medically expressed radiation damage does not occur below doses of 50 rads in the majority of exposed individuals. Exposure to patients and personnel from diag-

TABLE 8-2.—MEDICAL/DENTAL EXPOSURES TO THE POPULATION
(UNITED STATES)

TYPE PROCEDURE	NUMBER OF PERSONS EXPOSED	PERCENTAGE OF POPULATION
Radiographic	76 million	38.0
Fluoroscopic	9 million	5.0
Radiotherapeutic	400,000*	0.2
Dental radiographic	60 million	30.0

*This is an estimate of therapeutic procedures and does not include brachytherapy or teletherapy; therefore, this number is an underestimation.

nostic x-rays falls far below this dose, into the category of low doses—
the effects of which are classified as late effects and remain speculative and difficult to evaluate as discussed in Chapter 7. In addition to
the inherent difficulty in determining the effects of low doses, other
factors play a role in diagnostic radiology.

Exposure to patients is limited to one specific body area, rendering the determination of the systemic effects of these low doses
even more difficult. The added factor of chronic exposure to these low
doses must be considered in the case of occupationally exposed persons.

The primary biologic risk from doses of the magnitude received
in diagnostic radiology is probably the induction of chromosome
breaks in somatic and germ cells resulting in mutations and the health
impact of these mutations on the individual (including patients, personnel and fetus) and on future generations.

The Concept of Maximum Permissible Dose

Recognition of the harmful effects of radiation and the biologic
risks involved resulted in the establishment of limitations on the
amount of radiation received by all individuals including the general
population and occupationally exposed persons. Imposed in the
1920's, these limits have been constantly reduced to their current
levels (Table 8-3). Initially termed "tolerance doses," these limits are
now termed "maximum permissible doses" and have evolved from
the realization that there is no dose below which the risk of radiation
damage is nonexistent (i.e., no threshold dose). The philosophy be-
hind MPD's is that while there is no tolerance or safe dose of radiation, there is a dose below which there is assumed to be relatively

TABLE 8-3.—SOME MAXIMUM PERMISSIBLE WHOLE-BODY DOSES

YEAR	EXPOSURE OR DOSE
For occupationally exposed individuals:	
1925	0.1 of an erythema dose/year*
1934	0.1 R/day or 0.5 R/week
1949	0.3 rem/week or 15 rem/year
1956 to present	0.1 rem/week or 5.0 rem/year
For individuals in the general population:	
1952	0.03 rem/week
1958	5 rem/30 years
Present	0.170 rem/year

*Estimated to be approximately equivalent to 25 and 50 R/year from x-rays produced
by 100kv and 200kv potentials, respectively. From Dalrymple, G. V., *et al.: Medical
Radiation Biology* (Philadelphia: W. B. Saunders, 1973). Courtesy of M. E. Gaulden.

small biologic risk to occupationally exposed persons and to the general population. These risks, when weighed against other risks, are relatively minor and therefore acceptable, but it must be remembered that the higher the dose, the greater the risk. The reductions in MPD's directly reflect the increased knowledge, gained over many years, of the harmful effects of radiation. The drastic reduction in 1956 primarily reflects the increased knowledge of the ability of radiation to produce mutations in both somatic cells and, more importantly, in germ cells.

The present limits allow occupationally exposed individuals to receive a total body exposure of 5 rem/year after the age of 18. Because of the ability of cells to repair radiation damage, the accumulation of this limit is regulated on a quarterly basis; a total of 1.2 rem is allowed per calendar quarter. A single or cumulative exposure greater than the quarterly or yearly limit may constitute an overexposure necessitating investigation by the appropriate regulatory agency. However, occupationally exposed persons in diagnostic radiology rarely exceed their permissible limits of 5 rem/year, therefore accumulating a yearly "bank balance" of unused rem on which they may draw if an unusual situation should occur in which they are exposed to greater than the permissible limit. This amount is simply deducted from their "bank balance," allowing them a decreased rem account to draw upon. This account will again be rebuilt as unused rem "funds" are added.

In addition to limitations on total body exposures, limits also have been established for various organs, reflecting the radiosensitivity of the organ (Table 8-4). The rationale for MPD's can best be explained as an effort to keep the dose to all individuals as low as practicable. The benefits versus the risks of radiation exposure are constantly balanced with the exposure permitted *only* when the benefits exceed the biologic risks.

Risks in Diagnostic Radiology

Patients

Studies of circulating lymphocytes of patients at various times following diagnostic procedures have shown that the number and types of chromosome breaks increase with increasing dose. Skin exposures of 20 mR to 3 R produce no chromosome abnormalities in circulating lymphocytes. Doses of this magnitude are given during examinations such as chest radiographs and intravenous pyelograms (IVP's). Chromosome fragments have been observed in the lymphocytes of patients who have undergone cardiac catheterization, a procedure that results in exposures of 4 to 12 R and higher. Upper gastrointestinal (GI) and

TABLE 8-4.—CURRENT RECOMMENDED OCCUPATIONAL LIMITS
(NCRP)

TYPE OF EXPOSURE	DOSE CONDITION	RECOMMENDATION (REM)
Combined external and	Quarterly	3
internal whole-body	Annually (average)	5
	Annually (maximum)	10–15
	Long-term accumulated	
	to age N, beyond 18	5 (N–18)
Skin	Quarterly	—
	Annually	15
	Long-term accumulated	
	to age N, beyond 18	5 (N–18)
Thyroid and bone, feet and	Quarterly	—
ankles	Annually	15
Forearms	Quarterly	10
	Annually	30
Other organs	Quarterly	5
	Annually	15
Female workers of	Quarterly	—
reproductive capacity	Annually	2–3
	To fetus during	
	gestation period	0.5

barium enema (BE), procedures that may give doses up to 35 R*, produce both simple and complex aberrations (e.g., fragments, dicentrics and rings) in the chromosomes of circulating lymphocytes.

Although the health impact of these mutations in circulating lymphocytes is difficult to evaluate, the question certainly arises of the later development of leukemia. Some investigators feel that, particularly in procedures that result in higher doses, mutations probably have occurred in stem cells resulting in a continual propagation of these mutations in circulating lymphocytes. Whether these mutated cells eventually produce leukemia is unknown. However, studies of individuals with the disease have shown that one type of leukemia is associated with a specific chromosome abnormality, strongly suggesting a relationship between chromosome abnormalities and the development of this disease.

Though radiation is a known causative agent for leukemia (Chapter 7), the role of diagnostic radiology in the etiology of the disease remains speculative; at the present time, there is no *conclusive* evidence that doses from diagnostic examinations pose any hazard to adult patients in terms of leukemia. However, this does not rule out

---

*It is important to remember that exposures involving fluoroscopy vary according to the time involved. In addition, modern image intensifying systems greatly reduce fluoroscopic exposure.

the possibility that diagnostic doses may present other somatic hazards to the exposed individual.

Personnel

Occupationally exposed personnel present a different problem than do patients. In contrast to patients who are exposed to medical/dental radiation relatively few times during their lifetimes, personnel are exposed to chronic low doses throughout their professional lives. Chromosome mutations in circulating lymphocytes are much more frequent in occupationally exposed individuals, increasing with the number of years employed. Figure 8-1 shows the chromosomes from a cell of a radiologist whose lymphocytes had a number of chromosome mutations. Radiologists (the most thoroughly studied occupationally exposed group) at one time had a significantly increased incidence of leukemia as compared to other physicians. Today, however, this trend is decreasing; in addition, the increased incidence of other cancers in occupationally exposed individuals has disappeared. This decreased cancer incidence is directly attributable to the establishment of and reduction in the maximum permissible dose.

The rationale for MPD's is that, although there is no tolerance to radiation, there is an acceptable risk for occupationally exposed persons. The present limits are within this risk, because from the evidence available there does not appear to be excess danger to occupationally exposed personnel from radiation *provided*, of course, that all rules and regulations of radiation hygiene are followed.

Fetus

The damaging effects of radiation on the fetus are well known (Chapter 6) and have a direct relationship to diagnostic radiology. Although conclusive evidence implicates radiation in the production of many congenital abnormalities, the deforming effects on the fetus of doses from diagnostic procedures remain controversial, as does the evidence concerning the induction of leukemia (Chapter 7). Doses as low as 10 rads are believed to be harmful to the fetus during the first 6 weeks of gestation. If consulted, many radiobiologists will recommend a therapeutic abortion if a dose of 10 rads has been received at one time during the first 6 weeks in utero.

Although doses to the fetus from diagnostic procedures are difficult to estimate, Table 8-5 presents the estimated average fetal gonadal doses from various diagnostic procedures involving the abdomen. Chromosome mutations in the fetus as a result of irradiation in utero are presented in Table 8-6.

The major problem arising from exposure of the fetus to diagnostic x-rays is that the most radiosensitive period in the life of the

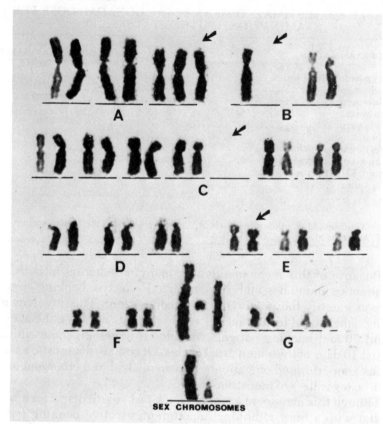

Fig. 8-1.—Chromosome damage in the lymphocyte of a long-practicing radiologist. *Arrows* indicate various types of damage. A dicentric chromosome and two acentric fragments are seen between *F* and *G*. There was a 7% incidence of chromosome mutations compared to 2% mutations in the non-occupationally exposed control. (From Dalrymple, G. V., *et al.*: *Medical Radiation Biology* [Philadelphia: W. B. Saunders, 1973], p. 75. Courtesy of M. E. Gaulden, Ph.D.)

fetus is the first 6 weeks when a woman is usually unaware of the pregnancy. As a result, it is strongly recommended that the possibility of pregnancy be ruled out on all female patients between the ages of 11 to 50 prior to diagnostic procedures and that the "10-day rule" be followed. This rule, suggested by the National Commission on Radiation Protection and Measurement (NCRP), recommends that all non-emergency diagnostic procedures involving the pelvis of women of childbearing age be conducted during the first 10 days of the menstrual cycle when the probability of pregnancy is very small.

TABLE 8-5.—ESTIMATED MEAN FETAL GONADAL DOSE PER EXAMINATION, UNITED STATES, 1970*

TYPE OF EXAMINATION	MRADS/EXAMINATION
Upper gastrointestinal series	
Radiographic	483
Fluoroscopic	170
Barium enema	
Radiographic	1140
Fluoroscopic	444
Cholecystography or cholangiogram	118
Intravenous or retrograde pyelogram	467
Abdomen, KUB, flat plate	153
Lumbar spine	658
Pelvis	353
Hip	206

*From Gaulden, M. E.: Possible effects of diagnostic x-rays on the human embryo and fetus, J. Arkansas Med. Soc., 70:424, 1974.

Because of this recommendation, many radiologists in both private practice and in hospitals have initiated "elective booking" procedures to identify the potentially pregnant patient. This involves obtaining a menstrual history on all female patients between the ages of 11 and 50 and booking patients considered nonemergencies during the first 10 days of their menstrual cycles. Of course, diagnostic examinations are performed on patients who are considered emergencies at any time regardless of menstrual cycle.

Though this may seem a cumbersome task, particularly for a large hospital with a busy radiology department, elective booking is feasible despite the workload of the department. Many hospitals have instituted this procedure. There are a number of methods by which this procedure may be done, either by the radiologist or by the referring physician. At the University Hospital of the Medical University of South Carolina, the questionnaire is part of the radiographic consultation form sent to the radiology department by the referring physician before the examination is performed, leaving to the physician's discretion the emergency status of the patient. Although this type of procedure alleviates the need for radiology department personnel to interview the patient and saves valuable time when a referring physician must be located because of the possibility of pregnancy in the patient, it does not afford the referring physician the knowledge of the radiologist concerning fetal doses or fetal effects of irradiation at the estimated time of gestation (factors that could influence a decision to either delay or go on with the procedure).

An alternate to this type of booking procedure is to interview the patient in the radiology department, thereby placing the responsi-

TABLE 8-6.—CHROMOSOME MUTATION IN FETUSES EXPOSED IN UTERO DURING DIAGNOSTIC RADIOGRAPHIC AND FLUOROSCOPIC PROCEDURES*

CASE NUMBER	ESTIMATED DOSE OR EXPOSURE	TYPE OF EXAMINATION	FETAL AGE AT EXPOSURE	TISSUE STUDIED	CHROMOSOME MUTATIONS
1	0.19 rad	Abdominal hysterosalpingography	10–11 weeks	—	None
2	3.0 R	—	6 weeks	Skin, lung	Translocations and others
3	3.9 rads	Barium meal, enema	6 weeks	Chorionic fragments	Deletions, dicentrics
4	—	"Radiopelvimetry"	1 week	Lymphocytes, fibroblasts	Extra chromosome markers

*From Dalrymple, G. V., et al.: Medical Radiation Biology (Philadelphia: W. B. Saunders, 1973). Courtesy of M. E. Gaulden.

bility of determining pregnancy status with the department. If a patient is found to be possibly pregnant, the referring physician is contacted before the procedure is initiated, leaving the decision concerning the necessity of the examination to his or her judgment. However, the radiologist can now estimate the fetal dose from the requested procedure and advise the referring physician of the potential dangers and risks involved, thus enabling the referring physician to make a more reasonable decision concerning the benefits to the patient versus the risks to the fetus from the procedure.

Two drawbacks to this latter type of booking are that the interviews may tie up a busy department and, more importantly, the patient may be caused unnecessary delay and discomfort if the referring physician must be located. These two limitations can be easily overcome and this type of booking system easily executed by thoughtful and careful planning, organization and administration of this elective booking procedure. Many radiology departments prefer this second booking system or a combination of both, i.e., the patient is briefly interviewed by the referring physician—this information as part of the consultation form is sent to the radiology department where an in-depth interview is conducted before initiation of the procedure.

Many different types of questionnaires may be used; the form used at the Medical University of South Carolina is shown in Figure 8-2. Examples of other types of questionnaires are shown in Figures 8-3, 8-4 and 8-5. Regardless of the questionnaire and procedure used,

Fig. 8-2.—Radiographic consultation form used at the Medical University of South Carolina Hospital. The elective booking section is printed in red on the form to catch the eye and remind the physician to elicit the required information from the patient before referral for the procedure.

MEDICAL UNIVERSITY HOSPITAL

DEPARTMENT OF RADIOLOGY

ROENTGENOGRAPHIC EXAMINATION OF:

PRIMARY CLINICAL DIAGNOSIS

THIS EXAM #

Previous X-rays		Films this exam	
☐ MUH	______	Fluoro	______
☐ Clinic	______	14x17	______
☐ Roper	______	11x14	______
☐ County	______	10x12	______
☐ Other	______	8x10	______

SPECIFIC INFORMATION SOUGHT FROM THIS STUDY

IF FEMALE BETWEEN AGES 11 TO 50, COMPLETE ONE OF THE FOLLOWING:

_____ 1. Patient is not considered pregnant.

_____ 2. Proceed with examination; benefits of study outweigh the risks.

Physician's Signature: ______________________

REPORT

University of Arkansas
Medical Center
University Hospital
Little Rock

RADIOLOGY DIAGNOSTIC REQUEST
(For THERAPY Use consultation form
For NUCL MED Use radio isotopes form)

IMPORTANT! IT MAY BE NECESSARY TO DELAY EXAM(S) IF INFORMATION IS INCOMPLETE

Requesting Service/Code | Hospital Location of Pt

UAMC XRAY Past 10 Yrs? YES ☐ NO ☐ | Pt. Cannot Walk; Need WHEELCHAIR ☐ STRETCHER ☐

Examination(s) Desired:

Use Identification Plate

Name

Unit No.   Race   Sex

Birth-
date

City

Purpose or Point of Concern (Optional)

IF FEMALE & LOWER ABDOMEN INCLUDED, COMPLETE THIS BLOCK:
➤ Patient is not considered pregnant because:

Date of Request

Exam is   Soon As ☐ or   Date
Desired: Possible

OR ⬇ (Signature of Interviewer)

Proceed with examination regardless of pregnancy or phase of menstrual cycle:
_____ M.D.

History

Full Signature

REQUESTED BY:
(Physician Authorization)

DIAGNOSIS: ☐Provisional?
                    ☐Proven?

Initials   Number

Fig. 8-3.—Request form used by the Diagnostic Roentgenology Section of the Arkansas Medical Center containing elective booking information in the lower right portion of the form. (From Dalrymple, G. V., et al.: *Medical Radiation Biology* [Philadelphia: W. B. Saunders, 1973], p. 302. Courtesy of G. V. Dalrymple, M.D. and M. L. Baker, Ph.D.)

elective booking should be followed by all diagnostic radiology departments, with the questionnaire then becoming a permanent part of the patient's record in both the department and in the hospital.

Future Generations

The harm to future generations is of primary concern in diagnostic radiology. Because radiation-induced mutations are nonthreshold, because they are usually considered to be detrimental and because they may be exhibited in generations long after exposure has occurred, the dose received by the gonads from diagnostic examinations are of major importance.

Table 8-7 lists doses to the gonads from various radiographic examinations for 1964 and 1970; fluoroscopy is not included.* As seen from this table, the doses received are far below the doubling dose (Chapter 7). In addition, they are not received by the whole population but only by that 65% of the general population exposed to x-rays for medical/dental purposes. For these reasons the doubling dose

---

*This information is taken from the most recent publication of the Bureau of Radiological Health, February 4, 1975.[36] At this time, information on fluoroscopy was unavailable.

```
┌────────────────────────────────────────────────────────────────────┐
│ FROM:  Department of Radiology                                       │
│                                                                      │
│ TO:    Referring Physician                                           │
│                                                                      │
│ RE:    Diagnostic Radiology Request (attached)                       │
│                                                                      │
│                                                                      │
│        ☐   Please complete information in general                    │
│                                                                      │
│        ☐   Please complete:_____         │
│                                                                      │
│        ☒   This patient may be pregnant. Either defer examination, or│
│            sign the statement, "Proceed with examination regardless  │
│            of pregnancy or phase of menstrual cycle."                │
│                                                                      │
│                                        Thank you for your cooperation.│
│                                        It helps us do a better job for│
│                                        you and your patient.         │
└────────────────────────────────────────────────────────────────────┘
```

Fig. 8-4.—Form used by the Diagnostic Roentgenology Section of the Arkansas Medical Center. This form is sent back to the ward if the section of the radiologic request form regarding pregnancy (Fig. 8-3) is not completed. (From Dalrymple, G. V., *et al.: Medical Radiation Biology* [Philadelphia: W. B. Saunders, 1973], p. 302. Courtesy of G. V. Dalrymple, M.D. and M. L. Baker, Ph.D.)

cannot be applied to determine the effects of these gonadal doses on future generations, although these doses may affect entire future populations.

To evaluate the genetic impact of these low doses on the whole population, the term *Genetic Significant Dose (GSD)* is used. This is an average figure calculated from the actual gonadal doses received by the exposed population, which also takes into account the expected contribution of these individuals to children in future generations. It is assumed that this dose, received by *every* member of the population, would have the same genetic effect as the doses that are now being received by the *proportion* of the population exposed to medical/dental x-rays.

Table 8-8 lists the GSD to the population from all types of radiation to which they are now exposed. As seen in the table, medical diagnostic radiation contributes the largest GSD to the population; the contribution from dental exposure is negligible. Radiotherapy accounts for only a small percentage because the majority of the individuals undergoing treatment are past childbearing age.

Previous reports indicated the GSD to be 55 mrads in 1964 and 36 mrads in 1970. However, a review of the data by the Bureau of Radiological Health (BRH) revealed errors, necessitating a new study and revision of the GSD. Based on the preliminary report recently issued by BRH,[36] the GSD from radiographic examinations *only* was 16

PROCEDURE BEFORE DOING DIAGNOSTIC RADIOLOGY EXAM

If all of the following:

- female between 11 and 50 years of age

- lower abdomen to be included in radiologic examination (e. g., Upper G. I., IVP, pelvis, lumbar spine, hysterosalpingogram, pelvic pneumogram)

- no statement from referring physician to go ahead in spite of pregnancy risk

- Patient is not an obvious emergency.

THEN ASK -- Is it possible that you are pregnant?

If "yes" - refer as follows:

   Outpatient - Attach "incomplete information" form to diagnostic request and have patient return with it to clinic.

   ER - Call referring physician.

   Inpatient - Have radiologist evaluate.

If "no" - document* reason as:

   has not started menstruating yet

   has stopped menstruating (menopause, hysterectomy, oophorectomy)

   bilateral tubal ligation

   taking birth control pills regularly

   has intra-uterine coil

   no intercourse since last period

   normal menstrual flow began less than 14 days ago

   other?

* make note on front of request
  sign it
  proceed with examination

Fig. 8-5.—Form completed by personnel in the Diagnostic Roentgenology Section of the Arkansas Medical Center as an additional screening for unsuspected pregnancy. (From Dalrymple, G. V., et al.: Medical Radiation Biology [Philadelphia: W. B. Saunders, 1973], p. 303. Courtesy of G. V. Dalrymple, M.D. and M. L. Baker, Ph.D.)

mrads in 1964 and 20 mrads in 1970 (Table 8-9). The percent contribution from various radiographic examinations to the 1970 GSD is presented in Table 8-10; examinations listed contributed 90% or more to the 1970 GSD. Although these figures do not include the contribution from fluoroscopy, previous studies indicate radiography contributed 96% to the GSD while fluoroscopy contributed only 4%. If this relationship holds true, the revised GSD figures may not change significantly when the fluoroscopic contribution is added.

TABLE 8-7.—ESTIMATED GONADAL DOSE FROM VARIOUS
RADIOGRAPHIC EXAMINATIONS BY SEX, UNITED STATES, 1964
AND 1970, NEW DOSE MODEL*

| | MRADS | | | |
| | 1964 | | 1970 | |
TYPE OF EXAMINATION	MALE	FEMALE	MALE	FEMALE
Skull	—	—	—	—
Cervical spine	—	—	—	—
Chest: radiographic	1	5	—	1
Chest: photofluorographic	—	5	2	3
Thoracic spine	46	17	3	11
Shoulder	—	—	—	—
Upper GI series	22	122	1	171
Barium enema	119	470	175	903
Cholecystography or cholangiogram	—	71	—	78
Intravenous or retrograde pyelogram	535	437	207	588
Abdomen, KUB, flat plate	63	248	97	221
Lumbar spine	108	507	218	721
Pelvis	443	119	364	210
Hip	718	196	600	124
Upper extremities	—	1	—	—
Lower extremities	38	—	15	—
Other abdominal examinations	296	213	857	524
All other	1	4	—	6

*From *X-Ray Exposure Study (XES); Revised Estimates of 1964 and 1970 Genetically Significant Dose,* Prerelease Report, (Rockville, Maryland: Bureau of Radiological Health, 1975).

Table 8-9 reveals that the contribution to the GSD from ovarian exposure doubled in 1970 as compared to 1964, while the male contribution decreased by almost one-half. Although the reasons for these changes have not been clearly defined, they may be related to two factors. First, between 1964 and 1970, all x-ray facilities showed a trend toward improved collimation (the mean ratio of beam area to film area for radiographs decreased approximately 30%), possibly resulting in exclusion of the testes from the primary beam as well as a reduction in scatter to this organ. Either or both of these possibilities would serve to reduce the GSD in males. On the other hand, there was approximately a 30% increase in the estimated mean skin exposure per film for anterior to posterior (AP) and posterior to anterior (PA) views of the abdomen. Because the ovaries are probably included in the majority of abdominal radiographic examinations, an increased skin exposure also would produce an increased ovarian exposure which would then be reflected as an increased GSD in females. The relationship of these and other factors to the increased ovarian and decreased testicular GSD awaits further investigation.

Of prime importance in these revised findings is that there is no

TABLE 8-8.—DOSES OF IONIZING RADIATION RECEIVED BY THE
HUMAN POPULATION

SOURCE	AVERAGE ANNUAL GSD (MREM)	
Natural background	125	
Medical uses:		
Diagnostic radiography	16	(1964 estimate)
	20	(1970 estimate)
Radiotherapy	3	
Nuclear medicine	0.2	
Occupational exposure (medical, dental, and atomic		
energy workers)	0.2	
Fallout from 1954–1962 period of nuclear weapons		
testing	1.5	
Miscellaneous (including television sets, luminous		
watches, cosmic radiation to high-flying aircraft,		
etc.)	2	
Nuclear power reactors (1970)	0.002	

statistically significant difference between the 1964 and the 1970
GSD, *despite* a 20% increase in the proportion of the population exposed. Although partially due to improvements in equipment, more
importantly this finding speaks highly for individuals in the profession and is a testimony to their interest in the patient's welfare.

The GSD is not an overall biologic indicator of radiation effects; it
does not evaluate the biologic effect of radiation on future generations
or indicate the number of mutations that will arise in future generations as a result of these doses. It only provides an estimate of the gonadal doses received from diagnostic radiation by the whole population. Its importance lies in the fact that, with a major segment of the
population involved, recessive mutations may accumulate and eventually be expressed in future generations, particularly because of the
mobility of our modern population and trends to marry from dissimilar
backgrounds.

TABLE 8-9.—ESTIMATED ANNUAL GSD CONTRIBUTIONS FROM
RADIOGRAPHIC EXAMINATIONS BY SEX, UNITED STATES, 1964
AND 1970, NEW DOSE MODEL*

	MRADS	
	1964	1970
Total	16	20
Male	8	5
Female	7	14
Fetus	1	1

*From *X-Ray Exposure Study (XES); Revised Estimates of 1964 and 1970 Geneti-
cally Significant Dose*, Prerelease Report (Rockville, Maryland: Bureau of Radiological
Health, 1975).

TABLE 8-10.—ESTIMATED ANNUAL GSD CONTRIBUTION FROM
RADIOGRAPHIC EXAMINATIONS, BY TYPE OF EXAMINATION,
UNITED STATES, 1970, NEW DOSE MODEL*

TYPE OF EXAMINATION	PERCENT CONTRIBUTION
Lumbar spine	20
Intravenous or retrograde pyelogram	16
Pelvis	12
Abdomen, KUB, flat plate	10
Barium enema	10
Hip	5
Other abdominal examinations	20
Radiographic examinations not listed	9

*From *X-Ray Exposure Study (XES); Revised Estimates of 1964 and 1970 Genetically Significant Dose*, Prerelease Report (Rockville, Maryland: Bureau of Radiological Health, 1975).

It is imperative that gonadal exposure be kept to a minimum during all diagnostic procedures and that all precautions be taken to ensure that the gonads receive *no* unnecessary exposure. Gonadal shields should always be used when the gonadal area is not of interest in the study; the use of shields decreases gonadal doses by 90%. The avoidance of unnecessary repeat exposures would decrease the GSD by 10%, and an additional 30% decrease would result from improved technic. In general a 50% decrease in GSD could be achieved with improved technic and education. It has been estimated that proper collimation alone (i.e., collimating the x-ray beam so that the beam is no larger than the film) would reduce the GSD by 65%!

Personnel and the Reduction of Patient Exposure

Both the radiologist and the radiologic technologist, as the administrators of radiation and guardians of radiation hygiene, can contribute greatly to reducing patient exposure and still obtain good quality radiographs through the following means:
1. Keep milliamperes and time of exposure to a minimum during fluoroscopic procedures.
2. Use the smallest field possible to obtain the necessary information.
3. Use proper filtration.
4. Use the proper technical factors, particularly KVP.
5. Avoid repeat films—the number of repeat films can be substantially reduced if the proper technical factors (KVP, MAS) are used during the first exposure—*remember*, repeat films add unnecessary exposure.
6. Avoid inclusion of the gonads in the primary beam.

7. Protect the testes with gonadal shields—their use decreases gonadal dose by 90%.
8. Set up the patient properly to reduce motion.
9. Determine that the patient is not pregnant.

Some of the above are the responsibility of the radiologist. However, many are directly dependent on the radiologic technologist's
knowledge and understanding of the work. The primary responsibility of the radiologic technologist includes patient positioning and
setting the technical factors on the x-ray unit for radiography and fluoroscopy. In essence, the technologist is the primary administrator of
radiation. Because of this, the technologist assumes a major role in
minimizing and controlling patient exposure and guarding radiation
hygiene. Thorough knowledge of the work contributes not only to
dose reduction but enhances the professional role of the technologist
by producing better quality radiographs from which information can
be derived by the radiologist.

Another aspect of patient dose reduction deals with the care and
feeding of x-ray machines and ancillary equipment. All equipment
must be in proper working condition to produce radiographs of good
quality. Screens should be periodically cleaned and proper screen
contact maintained with the film. Chemicals in the film processor
always should be fresh and the correct temperature always maintained in the processor. Neglect of these areas also can result in poor
quality films necessitating repeated examinations and unnecessary
patient exposure. Good technic and quality control in the x-ray room
can be easily negated by poor quality control in other areas of the department. A quality control program must be established and maintained in all areas of the radiology department and is the responsibility of all personnel.

Inherent in any effort to reduce patient exposure, whether inside
or outside the x-ray room, and in the establishment of a quality control
program throughout the department, is a reduction in personnel exposure. Repeated examinations cause unnecessary exposure to personnel as well as to patients; use of the smallest possible field size,
proper filtration and proper collimation not only enhance the study
and reduce patient exposure, but also reduce scattered radiation,
thereby reducing personnel exposure, particularly during fluoroscopy. Probably one of the most efficient means of reducing personnel
exposure is to reduce fluoroscopic time.

Although diagnostic radiology personnel rarely exceed the MPD,
special procedures (e.g., angiographic studies, particularly coronary
arteriography) by their nature greatly increase personnel exposure.
Therefore, in these areas careful consideration must be given to those

methods of dose reduction outlined in this chapter. In addition, the continually increasing use of radiation for diagnostic purposes may eventually be reflected in higher personnel exposure if methods of dose reduction and quality control programs are ignored. The practice of good radiation hygiene is important to the welfare of occupationally exposed personnel as well as of the patients.

The importance of personnel in reduction of exposure and its impact on both somatic and genetic effects might be better appreciated when considering the estimate of the Advisory Committee on the Biological Effects of Ionizing Radiations (BEIR) that 1,100 to 27,000 serious disabilities, congenital abnormalities, deaths, etc.; 1.2 to 12% of the overall ill health of the population of the United States; and 3,000 to 15,000 cancer deaths per year are the *somatic* risks arising in the general population from an exposure of 120 mrem/year —the MPD to the general population *excluding* medical/dental exposure!

Conclusions

The welfare of the patient is of primary concern to all individuals in the medical field. The major aim is to offer maximum assistance to the patient with as little adverse effect as possible. As a result, the benefits versus the risks of procedures are constantly being weighed. The use of diagnostic x-rays offers many benefits to the patient with minimal, if any, risk to individual health. However, the risk to future generations is not as easy to evaluate. For whatever reason, patients should always be exposed to as little radiation as possible.

All personnel in medical fields have a responsibility in the care of the patient and the minimizing of hazards. This is extremely important in diagnostic radiology, especially as the public is becoming more knowledgeable of and concerned with the genetic hazards of radiation. The lay press has contributed to public concern and has reminded us of our responsibilities to our patients. Concern by the patient, coupled with the attention of the lay press, poses medico-legal problems for the profession. An increased amount of litigation has been directed against the medical profession including radiology, particularly in the area of in utero exposure. Therefore, it is imperative from the standpoint of the patient's health, the health of personnel, the protection of future generations and the avoidance of medico-legal problems that only the best is offered to the patient and that greater attention is directed to technics of minimizing radiation hazards.

Good radiation hygiene is the responsibility of *every* individual in a radiology department. If these responsibilities are not assumed by individuals in all aspects of the profession, increasing federal regulations regarding x-ray usage will be enforced, thereby decreasing professional responsibilities.

REFERENCES

1. Antoku, S., and Russell, W. J.: Dose to active bone marrow, gonads, and skin from roentgenography and fluoroscopy, Radiology 101:669, 1971.
2. Brent, R. L., and Gorson, R. O.: Radiation exposure in pregnancy, in Moseley, R. D. (ed.): *Current Problems in Radiology* (Chicago: Year Book Medical Publishers, September-October 1972).
3. Busart, S.: Developer pH; its significance in quality control, Radiol. Technol. 45:413, 1974.
4. *Court-Brown, W. M., *et al.*: The incidence of leukemia following the exposure to diagnostic radiation in utero, Br. Med. J. 2:1599, 1960.
5. Dalrymple, G. V., *et al.*: *Medical Radiation Biology* (Philadelphia: W. B. Saunders, 1973).
6. *Effects on Populations of Exposure to Low Levels of Ionizing Radiation,* Report of the Advisory Committee on the Biological Effects of Ionizing Radiation, Division of Medical Sciences, National Academy of Science/National Research Council (Washington, D.C.: U.S. Government Printing Office, 1972).
7. Ford, D. D., *et al.*: Fetal exposure to diagnostic x-rays in leukemia and other malignant disease in childhood, J. Natl. Cancer Inst. 22:1093, 1959.
8. Gaulden, M. E.: Possible effects of diagnostic x-rays on the human embryo and fetus, J. Arkansas Med. Soc., 70:424, 1974.
9. Hammer-Jacobsen, E.: Therapeutic abortion on account of x-ray examination during pregnancy, Dan. Med. Bull. 6:113, 1959.
10. *Health Physics in the Healing Arts,* Proceedings of the Seventh Midyear Topical Symposium, San Juan, Puerto Rico, Document 11-14, 1972, DHEW Publication (FDA) 73-8029, March 1973.
11. MacMahon, B.: Prenatal x-ray exposure and childhood cancer, J. Natl. Cancer Inst. 28:1179–1191, 1962.
12. Margolis, A. R.: Lessons of radiobiology for diagnostic radiology, Calwell Lecture, 1972. Am. J. Roentgenol. Radium Ther. Nucl. Med. 117:741, 1973.
13. *Medical Radiation Information for Litigation,* Proceedings of a Conference, Baylor University College of Medicine, Houston, Texas, November 21–22, 1968, DMRE 69-3 (Washington, D.C.: U.S. Government Printing Office 1969).
14. Mole, R. H.: Radiation effects in man; Current views and prospects, Health Phys. 20:485, 1971.
15. Morgan, K. Z.: Biological effects of ionizing radiation; Lecture given at course "Environmental Analysis and Environmental Monitoring for Nuclear Power Generation," University of California, Berkeley, September 9–13, 1974.
16. Nader, R.: Wake up America; Unsafe x-rays! Ladies Home J., May 1968, p. 126.
17. Oliver, R.: 75 years of radiation protection, Br. J. Radiol. 46:854, 1973.

*Court-Brown, W. M., is occasionally indexed as Brown, W. M.

18. Penfil, R. L., and Brown, L. M.: Genetically significant dose to the U.S. population from diagnostic medical roentgenology, 1964, Radiology 90:209, 1968.
19. *Population Exposure to X-Rays, U.S., 1970*, DHEW Publication (FDA) 73-8047 (Washington, D.C.: U.S. Government Printing Office, 1973).
20. *Reduction of Radiation Dose in Diagnostic X-Ray Procedures*, Proceedings of a Symposium, Houston, Texas, July 1971, DHEW Publication (FDA) 73-8009 (Washington, D.C.: U.S. Government Printing Office, 1972).
21. *A Review of Determination of Radiation Dose to the Active Bone Marrow from Diagnostic X-Ray Examinations*, DHEW Publication (FDA) 74-8007 (Washington, D.C.: U.S. Government Printing Office, 1973).
22. Rugh, R.: *From Conception to Birth: The Drama of Life's Beginnings* (New York: Harper & Row, 1971).
23. Rugh, R.: Why Radiobiology? Radiology 82:917, 1964.
24. Sagan, L. A.: Human effects of low level radiation; a critique, Proceedings of the Ninth Hannford Biology Symposium, 1969, AEC Symposium Series 17, pp. 719–730.
25. Stein, J. J.: The carcinogenic hazards of ionizing radiation in diagnostic and therapeutic radiology, Cancer 17:278–287, 1967.
26. Sternglass, E. J.: Evidence for low level radiation effects on the human embryo and fetus, Proceedings of the Ninth Hannford Biology Symposium, 1969, AEC Symposium Series 17, pp. 651–660.
27. Stewart, A., and Kneale, G. W.: Radiation dose effects in relation to obstetric x-rays and childhood cancers, Lancet 1:1185, 1970.
28. Stewart, A., *et al.*: A survey of childhood malignancies, Br. Med. J. 1:1495, 1958.
29. Stewart, A., *et al.*: Malignant disease in childhood and diagnostic irradiation in utero, Lancet 2:447, 1956.
30. Stewart, A.: *An Epidemiologist Takes a Look at Radiation Risks*, DHEW Publication (FDA) 73-8024, January 1973.
31. Stone, R. S.: Common sense in radiation protection applied to clinical practice, Am. J. Roentgenol. Radium Ther. Nucl. Med. 78:993, 1957.
32. Stone, R. S.: Concept of maximum permissible exposure, Carmen lecture, Radiology 58:639, 1952.
33. Warshofsky, F.: Warning—X-rays may be dangerous to your health! Readers' Dig., August 1972, p. 173.
34. Whaley, M. H.: Clinical dosimetry during the angiographic examination, Radiology 150:627, 1974.
35. *X-Ray Examinations: A Guide to Good Practice*, ACR Commission on Radiologic Units, Standards and Protection and the Bureau of Radiological Health (Washington, D.C.: U.S. Government Printing Office, 1971).
36. *X-Ray Exposure Study (XES); Revised Estimates of 1964 and 1970 Genetically Significant Dose*, Pre-Release Report (Rockville, Maryland: Bureau of Radiological Health, 1975).

9/Clinical Radiobiology II: Nuclear Medicine

Radiopharmaceuticals have been used in the diagnosis and treatment of disease since 1946 and, therefore, are relative newcomers to the medical uses of radiation. Today the specialty of nuclear medicine is a dynamic, growing and vital area of medicine with new radiopharmaceuticals and diverse applications of these materials being developed constantly as well as new and improved instrumentation.

In 1966 a survey estimated that approximately 1.5 million people were exposed to radiopharmaceuticals through the following procedures:

in vivo function studies	55%
scanning	42%
therapy	3%

At that time, it was estimated that a 15–20% increase in radiopharmaceutical use was occurring yearly. More recent studies have projected this figure to 25–30%. Based on these figures and a projected sevenfold increase in radiopharmaceutical use in the 1970's, approximately 6 million persons were administered radiopharmaceuticals in 1975; this figure is probably a conservative estimate.

Although the percentage of the population exposed to radiation via nuclear medicine procedures is low in comparison to diagnostic radiology, the health impact of these procedures is still of concern. As in diagnostic radiology, nuclear medicine procedures result in low doses, the potential biologic risks of which probably are carcinogenesis (particularly leukemia) and genetic effects. Those groups at risk include patients, personnel, fetuses and future generations.

Patient Exposure

Both the diagnostic and therapeutic uses of radiopharmaceuticals are dependent on the accumulation of the material in the target organ, i.e., the organ of interest. Some radiopharmaceuticals have an affinity for certain organs that are not necessarily the organ of interest; these organs are termed "critical organs." The critical organ is that organ which, although it may not accumulate the greatest percentage of the

171

radiopharmaceutical, dictates the amount of a radiopharmaceutical that can be administered. This amount is based not only on the accumulation in the organ but, more importantly, on the radiation sensitivity of the organ. The critical organ is therefore the dose-limiting factor in a radiopharmaceutical procedure. Because the determination of biologic effect and health risk to the patient is dependent on the amount of radiation received, the critical organ is of primary importance. However, radiopharmaceuticals are transported throughout the body by the bloodstream resulting in exposure to the total body, which is also of concern. Table 9-1 lists the exposure to critical organs and total body from widely used radiopharmaceuticals.

Two general factors are considered when estimating patient exposure in nuclear medicine: physical and biologic variables. The physical factors, e.g., the type of radiation emitted, energy and physical half-life, are well defined. The biologic variables are less well defined and, unfortunately, play a significant role in determining radiation exposure to the individual, thereby contributing greatly to uncertainties in exposure estimations. These biologic variables result in problems in determining exposure in nuclear medicine, particularly in terms of the patient, unique from those encountered in other medical uses of radiation.

Biologic Variables

One biologic factor that plays a large role is the fact that most patients undergoing nuclear medicine procedures are ill. The critical organ values listed in Table 9-1 represent exposure to healthy individuals. These figures, which may vary widely even among healthy persons, may be changed to an even greater extent by the presence of disease in the organ. Disease may affect the size or the function of the organ, thus influencing the amount of radionuclide accumulated, which in turn alters the radiation exposure to the critical organ. In addition, exposure to secondary organs that accumulate the radionuclide may be changed as well.

A second major problem is the age of the patient. Physiologic processes, metabolic rates and organ sizes in children differ from those in adults and vary even between infants and children (Table 9-2). Any one or a combination of these factors can result in an alteration in the uptake of radionuclide in the body and the exposure to various organs. Table 9-3 presents the doses received by organs in children from various radionuclides. Note the change in dose from each radionuclide as a function of age; in all cases, dose decreases with increasing age. Although lower doses are given to children undergoing nuclear medicine procedures, many questions remain concerning the fate of these materials in the body. Unlike adults, where a "standard" or "refer-

TABLE 9-1.—RADIATION DOSE FROM COMMONLY USED
RADIOPHARMACEUTICALS*

PROCEDURE AND AGENTS	USUAL ADMINISTRATION DOSE (mCi)	RADIATION DOSE (RADS)		
		TARGET ORGAN		WHOLE-BODY
A. Brain scan:				
^{203}Hg-chlormerodrin	0.7–0.9	Kidney**	70–90	1.2
^{197}Hg-chlormerodrin	0.7–1.0	Kidney	8–10	0.083
^{99m}Tc-pertechnetate	5.0–10.0	Colon	1–2	0.2
^{113m}In-DTPA	5.0–10.0	Bladder	2.5–5	0.05–0.15
B. ^{131}I-HSA cisternography				
normal	0.1	Spinal cord	7.2	0.05–0.1
hydrocephalic	0.1	Spinal cord	12.3	0.05–0.1
cervical block	0.1	Spinal cord	58.7	0.05–0.1
C. Lung scan				
^{131}I-MAA	0.3	Lung	1–3	0.12
^{99m}Tc-MAA	1.0–3.0	Lung	0.4–1	0.01
^{99m}Tc-albumin microspheres	1.0–3.0	Lung	0.4–1	0.01
^{113m}In-Fe (OH)$_3$ particles	1.0–3.0	Lung	0.75–2	0.012–0.036
^{133}Xe	5.0–10.0	Lung	0.25–0.5	0.001–0.002
D. Cardiovascular blood pool				
^{131}I-HSA	0.2–0.3	Blood	2.9–5	0.2–0.4
^{99m}Tc-HSA	1.0–3.0	Blood	0.04–0.12	0.01–0.03
^{113m}In-transferrin	1.0–3.0	Blood	0.04–0.12	0.01
E. Thyroid scan				
^{131}I	0.05	Thyroid	65–90	0.2
^{125}I	0.05–0.1	Thyroid	45–90	0.06
^{123}I	0.05–0.1	Thyroid	1–2	0.003
^{99m}Tc-pertechnetate	1.0	Thyroid	0.2	0.01
F. Liver scan				
^{198}Au-colloid	0.1–0.15	Liver	4–8	0.1–0.25
^{99m}Tc-sulfur colloid	1–3	Liver	0.3–1	0.008–0.02
^{113m}In-colloid	1–3	Liver	0.5–1	0.015–0.03
^{131}I-Rose Bengal	0.15–0.3	Liver	0.2–1.4	0.2–0.4
G. Spleen scan				
^{99m}Tc-sulfur colloid	1–3	Liver	0.3–1	0.008–0.03
^{113m}In-colloid	1–3	Liver	0.5–1	0.015–0.03
^{51}Cr-heated RBC's	0.1–0.3	Spleen	4–10	0.05–0.07
H. Pancreas scan				
^{75}Se-selenomethionine	0.25	Pancreas	3.5	
		Liver	7	
		Gonads	1.3–2.6	0.9–2.5
I. Bone scan				
^{85}Sr	0.1	Bone	3.1–4.6	0.68–1.6
^{87m}Sr	1–3	Bone	0.1–0.5	0.02–0.06
^{18}F	1–2	Bone	0.12–0.4	0.03–0.07
J. Kidney scan				
^{197}Hg-chlormerodrin	0.1–0.15	Kidney	1.2–1.8	0.01–0.02
^{99m}Tc-iron ascorbate	1–2	Kidney	0.5–1	0.008
^{99m}Tc-DTPA	1–2	Kidney	0.05–0.1	0.03
^{131}I-orthoiodohippurate	0.2–0.4	Kidney	0.2–0.4	0.006–0.012

*PDR for Radiology and Nuclear Medicine, Physician's Desk Reference (Oradell,
New Jersey: Medical Economics Company, 1972).
**May be reduced 40 to 50 % by a prior blocking dose of nonradioactive chlormerodrin.

TABLE 9-2.—BODY WEIGHTS AND ORGAN WEIGHTS FOR VARIOUS AGES*

ORGAN	NEWBORN	1 YEAR	WEIGHT (G) 5 YEARS	10 YEARS	15 YEARS	ADULT
Whole-body	3,540.0	12,100.0	20,300.0	33,500.0	55,000.0	70,000
Thyroid	1.9	2.5	6.1	8.7	15.8	20
Kidney	23.0	72.0	112.0	187.0	247.0	300
Liver	136.0	333.0	591.0	918.0	1,289.0	1,700
Spleen	9.4	31.0	54.0	101.0	138.0	150

*Kereiakes, J., et al.: Patient and personnel dose during radioisotope procedures, in Medical Radiation Information for Litigation, Proceedings of a Conference, Baylor University College of Medicine, Houston, Texas, November 21-22, 1968, DMRE 69-3 (Washington, D.C.: U.S. Government Printing Office, 1969).

ence" person is used for exposure calculations, a "standard" child has not been established. In all likelihood, based on the information in Tables 9-2 and 9-3, there will have to be at least two standards for children: one for infants and one for older children.

The biologic effect of a radionuclide is dependent on both its physical and biologic half-life, the latter being a function of the rate at which the nuclide is metabolized and eliminated from the body. Less is known concerning the biologic half-lives than the well-defined physical half-lives. Unfortunately, the biologic half-life is the main determinant of radiation exposure in an organ, because this factor determines the amount of time the radiopharmaceutical remains in the organ.

Although dose estimates in nuclear medicine are sometimes difficult, more than 30 years' experience reveals no significant adverse effects from these procedures. In general, diagnostic nuclear medicine procedures are believed to pose no more hazard to the patient than are diagnostic radiologic procedures and are not considered to deliver excessive doses. In fact, patient exposures from nuclear medicine are usually lower than those from radiography. In addition, the dose rate from radionuclides is lower than from diagnostic x-rays, resulting in less damage because of the potential for repair. From the evidence available, the risk of inducing late effects (particularly leukemia) in adult patients from diagnostic nuclear medicine procedures is very small—on the same order as the risk of leukemia induction from radiography.

The risks to children from diagnostic nuclear medicine procedures are more difficult to evaluate. Based on the fact that children are generally more sensitive to radiation than are adults and that many physiologic and metabolic factors concerning radionuclide use in children are unknown, doses from diagnostic procedures may pose more of a hazard to children than to adults.

TABLE 9-3.—RADIATION DOSE FROM RADIOPHARMACEUTICALS
IN CHILDREN*

RADIONUCLIDE USED	AGE	EFFECTIVE HALF-LIFE (DAYS)	DOSE TO CRITICAL ORGAN (MRADS/μCi)	
^{131}I	2 days	4.7	32	Thyroid
	1 month	7.0	10–32	Thyroid
	4 years	6.3	4.3–10	Thyroid
	15 years	5.9	1.7	Thyroid
^{51}Cr-sodium chromate	4 months	15.0	4.5	Whole-body
	5 years	20.0	1.0	Whole-body
^{59}Fe-ferrous citrate	5 years	38.0	65	Whole-body
	15 years	39.0	29	Whole-body
^{197}Hg-chlormerodrin	3 years	2.6	0.16	Whole-body
			68.1	Kidney
	12 years	2.6	6.07	Whole-body
			39.0	Kidney
^{85}Sr-strontium nitrate	4 years	58.0	16.3	Whole-body
			68.3	Bone
	12 years	53.0	8.6	Whole-body
			32.8	Bone
	18 years	50.0	4.7	Whole-body
			27.0	Bone

*Kereiakes, J., et al.: Patient and personnel dose during radioisotope procedures, in Medical Radiation Information for Litigation, Proceedings of a Conference, Baylor University College of Medicine, Houston, Texas, November 21–22, 1968, DMRE 69-3 (Washington, D.C.: U.S. Government Printing Office, 1969).

Dose Reduction

In keeping with today's trend of minimizing exposures, particularly because more persons are medically exposed to radiation including diagnostic nuclear medicine procedures, methods of dose reduction always are being sought.

There are a number of methods presently available to reduce exposure. The type of radiation emitted by the radionuclide plays a primary role. Imaging is dependent on the detection of γ-rays outside the body; radionuclides that emit β, low-energy x-rays, or conversion electrons do not contribute to the study but only to the radiation exposure of the patient. As a result, radionuclides that are pure γ-emitters or have a very small amount of low-energy radiation are used. This is the reason why technetium-99m (^{99m}Tc) has gained such widespread use.

A second method of dose reduction is the use of either stable agents or agents chemically similar to the radionuclide that will selectively block and decrease uptake in critical organs *not* of interest in the study (e.g., Lugol's solution to reduce thyroid uptake of radioactive iodine; Mercuhydrin to reduce kidney uptake of mercury-labeled neohydrin). In addition to reducing unnecessary irradiation to the critical organ, this method also enhances uptake in the organ of interest by increasing the available amount of radionuclide.

Future developments on reducing exposure include improvements in instrumentation that possibly can reduce patient exposure by a factor of ten. The trend toward the use of radionuclides with short half-lives is another method of reducing exposure; however, in this case the problem of dose rate appears. Radiobiologic evidence indicates that low-dose rates are biologically less damaging than high-dose rates, possibly because low-dose rates allow for more cellular repair to occur. The question eventually must be answered as to whether short-lived radionuclides that give small absorbed doses at high-dose rates are less hazardous than long-lived radionuclides that result in larger doses but at lower dose rates. Short-lived radionuclides are one answer to dose reduction; they are not the sole solution. Probably the greatest contribution to dose reduction can be made by the use of nuclides with shorter effective half-lives—which is dependent on greater knowledge of biologic half-life.

Although many of the above suggestions are outside the realm of the nuclear medicine technologist, the technologist *can* and *should* make a major contribution to reducing patient exposures. A sufficient amount of radionuclide must be given to the patient to obtain the desired results. An insufficient amount may be responsible for failure to define pathology in the organ, in which case the patient is "protected out of a diagnosis."* In addition, a repeat examination may be ordered which, in the long run, results in a higher patient exposure than if the correct amount had been initially administered.

In either case, the nuclear medicine technologist can greatly contribute to dose reduction by ensuring that:

1. The correct amount of radiopharmaceutical is always administered to the patient.
2. Instruments are calibrated to ensure that the amount administered is accurate.
3. All equipment is checked on a routine basis to ensure proper functioning.

In essence, a functional quality control program is a necessity in nuclear medicine.

The role of the nuclear medicine technologist is of particular importance in dose reduction when considering the estimate that nuclear medicine procedures may contribute 15% of the total somatic dose of all individuals by the year 2000. One institution has reported an increased average whole-body dose to patients from 100 mR in

*Bender, M. A., in *Reduction of Radiation Exposure in Nuclear Medicine*, p. 145.[14]

1964, to 160 mR in 1968. This increase was directly related to an increased use of radiopharmaceuticals in this time interval, a trend that continues today. Quality control assumes particular importance in view of these trends.

Personnel

Recent surveys of nuclear medicine personnel have revealed that exposure of these persons is rising. This increase in personnel exposure is directly attributable to the increased use of high-activity generators in nuclear medicine laboratories.

A recent survey revealed that ^{99m}Tc is the most widely used radionuclide for scanning purposes (Table 9-4). This same survey observed that ^{99m}Tc is the single largest contributor to hand exposure of nuclear medicine personnel, which is due primarily to handling of the radioactive material, *not* of the patients. It has been estimated that 95% of hand exposure is delivered during preparation and injection of ^{99m}Tc, 5% during assay and a negligible dose during elution. Although minimal exposure occurs during the daily handling and preparation of the technetium, cumulative exposure could be excessive (i.e., exceed the maximum permissible dose). Table 9-5 presents a summary of

TABLE 9-4.—RADIOPHARMACEUTICALS ADMINISTERED AT 69
HOSPITALS (1971)*

ORGAN	RADIOPHARMACEUTICAL	PERCENT OF TOTAL
Brain	^{99m}TcO$_4$Na	100.0
Liver	^{99m}Tc-S-colloid	90.0
	^{198}Au-colloid	8.5
	^{131}I-Rose Bengal	1.5
Lung	^{131}I-MA-albumin	84.7
	^{99m}Tc-ferrous hydroxide	6.4
	^{99m}Tc-MA-albumin	5.5
	^{99m}Tc-albumin microspheres	3.4
Thyroid	^{131}INa	76.6
	^{99m}TcO$_4$Na	21.9
	^{123}INa	1.5
Kidney	^{99m}Tc-DTPA	45.0
	^{197}Hg-chlormerodrin	30.8
	^{131}I-hippurate	12.0
	^{203}Hg-chlormerodrin	12.0
Bone	^{85}Sr-chloride	61.3
	^{18}FNa	38.7
Pancreas	^{75}Se-methionine	100.0

*From Lombardi, Max H., *et al.*: Survey of radiopharmaceutical use and safety in 69 hospitals, in *Health Physics in the Healing Arts*, Proceedings of the Seventh Midyear Topical Symposium, San Juan, December 1972, DHEW Publication (FDA) 73-8029, March, 1973.

exposure to the total body, hands and eyes of nuclear medicine personnel from ^{99m}Tc.

Estimates of hand and total body exposure to nuclear medicine technologists from technetium range from 160 to 200 mR/week for hand exposure and from 40 to 70 mR/week for total body exposure. The average hand exposure per administration of ^{99m}Tc is 11 mrem. On an annual basis, the cumulative hand exposure could be 15 rem, one-fifth of the permissible hand dose (75 rem/year), depending on the number of procedures performed. Although these exposures are below established MPD's, the increased use of radiopharmaceuticals, particularly ^{99m}Tc, could contribute to higher cumulative hand exposures and a resultant increased possibility of radiation injury.

A possible solution to this problem is to reevaluate radiation safety practices in nuclear medicine. Throughout the years, two systems of safety procedures developed for radionuclides, based on the amount used. Diagnostic doses were generally low-level activities, i.e., microcurie amounts, while therapeutic doses were in the millicurie range. However, the advent of high-activity generators resulted in the use of millicurie amounts for diagnostic purposes but the handling procedures remain based on lower activities. Adjustment of handling procedures to reflect these higher activities could possibly reduce personnel exposures.

An efficient and feasible method of reducing hand exposure is by routine use of lead syringe shields. These shields reduce the dose-rate

TABLE 9-5.—AVERAGE EXPOSURE FROM ^{99m}Tc TO VARIOUS BODY
AREAS OF TECHNOLOGISTS*

AREA	AVERAGE mR/WEEK		AVERAGE R/YEAR
Body	12.6		0.7
	9.0		0.5
	20.2	(Range)	1.1
	14.1	(3–80)	0.8
Hand	79.6		4.2
	51.7		2.7
	43.3	(Range)	2.2
	62.9	(10–380)	3.3
Eyes	28.3		1.5
	10.0		0.5
	25.0	(Range)	1.3
	24.0	(3–35)	1.3

*From Mayes, M. G., et al.: Study of radiation exposure from technetium generators at three hospitals, in Health Physics in the Healing Arts, Proceedings of the Seventh Midyear Topical Symposium, San Juan, December 1972, DHEW Publication (FDA) 73-8029, March 1973.

at the surface of a syringe by a factor of 300 resulting in a tenfold reduction in hand exposure—from 15 to 1.5 rem/year. The hand exposure of 6 nuclear medicine technologists is presented in Table 9-6; note that Technologist 3 had the lowest hand exposure although this individual performed the most procedures using ^{99m}Tc (118 procedures versus 4—the least number). This greatly reduced hand exposure was due to the consistent use of a syringe shield.

Methods of good radiation hygiene always should be practiced by personnel, particularly when these methods are easy and feasible. Nonuse of syringe shields constitutes both poor technic and poor radiation hygiene.

Another method of reducing personnel exposure is to rotate the duties of milking the generator and administration of ^{99m}Tc. In addition, regardless of the radionuclide being used, the technologist should always work in front of a shield to reduce total body and gonadal exposures. Quality control applies to personnel as well as to patient protection.

Fetus

Protection of the fetus is also of concern in nuclear medicine, as it is in radiography. Many radionuclides, e.g., ^{131}I and ^{99m}Tc, cross the placental barrier and localize in fetal tissues. Table 9-7 presents the dose to the fetus from ^{197}Hg, and Table 9-8 shows fetal dose from various radiopharmaceuticals used in placental localization. Because of the radiosensitivity of fetal tissues, biologic damage could occur.

TABLE 9-6.—HAND EXPOSURE TO SIX NUCLEAR MEDICINE
TECHNOLOGISTS FROM ^{99m}Tc*

TECHNOLOGIST	INTERVAL (DAYS)	NUMBER ^{99m}Tc PROCEDURES	SYRINGE SHIELD	TOTAL DOSE (MREM)**		
				THUMB	INDEX	RING
1	14	4	No	940	1750	135
2	24	46	No	1350	1680	460
3	13	118	Yes	150	235	120
4	5	60	No	660	710	500
5	3	85	No	230	360	180
6	10	48	No	830	780	540
Totals	69	361		4160	5515	1935
Mrem/administration				11.5	15.3	5.4
Average hand dose Mrem/administration				10.7		

*Lombardi, M. H., *et al.*: Survey of radiopharmaceutical use and safety in 69 hospitals, in *Health Physics in the Healing Arts*, Proceedings of the Seventh Midyear Topical Symposium, San Juan, December 1972, DHEW Publication (FDA) 73-8029, March 1973.
**Each value is the mean of two dosimeters.

TABLE 9-7.—FETAL UPTAKE OF ^{197}Hg*

ORGAN	WHOLE ORGAN WEIGHT (G)	PERCENT ^{197}Hg/g ($\times 10^{-4}$)	CALCULATED ABSORBED DOSE TO FETUS (RADS)
Heart	4.6	7.3	1.2
Lung	7.3	11.9	3.7
Liver	18.0	19.5	6.1
Maternal uterine	—	8.5	—
Placenta	—	25.2	—
Umbilical cord	—	9.6	—

*Uptake by a 25-cm (crown to rump) fetus 24 hours after the mother, scheduled for therapeutic abortion, was given 200 μCi ^{197}Hg-chlormerodrin intravenously. (From Sy, W. M., *et al.*: Radiation dose in a human fetus following use of ^{197}Hg, Radiology 103:139, 1972.)

Elective booking procedures such as those used in diagnostic radiology should be routinely used to identify the potentially pregnant patient in nuclear medicine. Figure 9-1 presents a form used by the Arkansas Medical Center for obtaining elective booking information in nuclear medicine.

Future Generations

Gonadal doses from various radiopharmaceuticals are listed in Table 9-9. In 1970 the GSD from nuclear medicine was estimated to be 0.3 mR; compared with the GSD from diagnostic radiology (20 mR), this amount is relatively minor. This low contribution to the GSD is due to two factors. Until 1970 only a small proportion of the general population was exposed to nuclear medicine procedures. The estimated 20% increase in the use of radiopharmaceuticals for diagnostic purposes will result in exposure of a larger number of individuals. In addition, in the past a large segment of the potential childbearing individuals—the genetically significant population—were not exposed to nuclear medicine procedures because of the warning that

TABLE 9-8.—DOSE TO FETUS DURING PLACENTAL LOCALIZATION*

RADIOPHARMACEUTICAL	USUAL ADMINISTRATION DOSE (mCi)	TARGET ORGAN	RADIATION DOSE (RADS)	WHOLE-BODY
^{131}I-HSA	0.005–0.010	Blood	0.005	0.004
^{99m}Tc-albumin	1.0	Blood	0.01	0.01
^{99m}Tc-pertechnetate	0.5–1.0	Colon	—	0.03
^{113m}In-transferrin	1.0	Blood	0.008	0.008

PDR for Radiology and Nuclear Medicine, Physician's Desk Reference (Oradell, New Jersey: Medical Economics Company, 1972).

FORM FOR PATIENT INTERVIEW

FOR FEMALE PATIENTS

YOU ARE BEING EXAMINED IN NUCLEAR MEDICINE AT THE REQUEST OF YOUR PHYSICIAN. THIS REQUIRES THE USE OF MATERIALS WHICH CONTAIN SMALL AMOUNTS OF RADIO-ACTIVITY. BECAUSE OF A POSSIBLE SMALL RISK TO UNBORN CHILDREN. IT IS NECESSARY FOR US TO DETERMINE IF YOU POSSIBLE COULD BE PREGNANT. WE WOULD APPRECIATE YOU ANSWERING THESE QUESTIONS HONESTLY SO WE MAY PROCEED SAFELY FOR YOUR BENEFIT. SOME OF THE QUESTIONS MAY NOT APPLY TO YOU, AND IF NOT, PLEASE WRITE "NO" IN THE BLANK.

DATE OF LAST MENSTRUAL PERIOD _____.

ARE YOU, OR COULD YOU BE PREGNANT NOW? YES_____ NO_____

DO YOU TAKE BIRTH CONTROL PILLS NOW? YES_____ NO_____

DO YOU USE AN INTRAUTERINE CONTRACEPTIVE DEVICE (LIKE A COIL)

NOW? YES_____ NO_____

HAVE YOU HAD AN OPERATION WHICH HAS MADE YOU STERILE? YES_____ NO_____

SIGNATURE_____

RADIOISOTOPE ADMINISTERED BY _____, M. D.

INSTRUCTIONS FOR NUCLEAR MEDICINE PERSONNEL

FOR FEMALE PATIENTS 11-50 YEARS OLD

PROCEED WITH TEST IF:

(1) MENSTRUAL PERIOD BEGAN LESS THAN 14 DAYS AGO

(2) "YES" ANSWERS TO ANY OF LAST 3 QUESTIONS

CHECK WITH PHYSICIAN IF:

(1) DATE OF LAST MENSTRUAL PERIOD IS OVER 14 DAYS

FROM DATE OF EXAMINATION

(2) PATIENT COULD BE PREGNANT

(3) NO "YES" ANSWERS TO LAST 3 QUESTIONS AND (1) ABOVE

APPLIES

Fig. 9-1.—Elective booking form used by the Nuclear Medicine Section at the Arkansas Medical Center. (From Dalrymple, G. V., *et al.: Medical Radiation Biology* [Philadelphia: W. B. Saunders, 1973]. Courtesy of G. V. Dalrymple, M.D. and M. L. Baker, Ph.D.)

radiopharmaceuticals were absolutely contraindicated in persons under 18 years of age. This factor also is changing as larger numbers of children are receiving radiopharmaceuticals for diagnostic purposes.

Because of the changing trends in the administration of radionuclides to all persons, particularly the increased use in persons under the age of 18, nuclear medicine will contribute a larger percentage to the GSD in the next few years. Although the impact of these changes

TABLE 9-9.—GONADAL DOSE FROM VARIOUS RADIONUCLIDES*

RADIONUCLIDE USED	PROCEDURE PERFORMED	ACTIVITY ADMINISTRATION (mCi)	DOSE ESTIMATE(RADS)		
			TOTAL BODY	GONADS MALE	FEMALE
99mTc-colloid	Liver scan	3.0	0.05	0.04	0.07
99mTc-sulfur colloid	Bone marrow scan	5.0	0.08	0.06	0.11
198Au-colloid	Bone marrow scan	2.5	5.8	0.98	1.5
199Au-colloid	Bone marrow scan	2.5	1.7	0.28	0.49

*Kereiakes, J., *et al.*: Patient and personnel dose during radioisotope procedures, in *Medical Radiation Information for Litigation*, Proceedings of a Conference, Baylor University College of Medicine, Houston, Texas, November 1968, DMRE 69-3 (Washington, D.C.: U.S. Government Printing Office, 1969).

on future generations remains to be seen, the increased contribution to the GSD from nuclear medicine certainly will contribute to the estimate by the International Commission on Radiation Protection (ICRP) that medical radiation exposure may introduce into future generations approximately 30,000 deaths from malignancies, stillbirths and spontaneous abortions because of genetic damage.

Therapeutic Use of Radionuclides

In 1966, 44,000 persons received radiopharmaceuticals for therapeutic purposes. The diseases most often treated by this method include hyperthyroidism, thyroid carcinoma and polycythemia vera. The radiopharmaceuticals used are iodine-131 (^{131}I) and phosphorus-32 (^{32}P).

Hyperthyroidism

The use of ^{131}I in the treatment of hyperthyroidism has been a successful means of controlling this disease for more than 20 years. However, the question of an increased incidence of leukemia and thyroid carcinoma in individuals treated remains controversial.

A study of patients treated for hyperthyroidism with ^{131}I revealed a slight increase of leukemia in this population when compared to the general population; however, hyperthyroid patients treated by surgical means showed this same increased incidence. Therefore, the induction of leukemia following ^{131}I treatment for hyperthyroidism appears improbable although certainly not impossible.

The question also arises concerning the incidence of thyroid carcinoma in hyperthyroid patients following ^{131}I therapy. Irradiation of the thyroid can certainly result in cancer as evidenced by the increased incidence of this disease in children irradiated for enlarged thymus glands and the Marshall Islanders exposed to fallout radiation.

Previously, this treatment method was not used on persons below 40 years of age; this age limit now has been recently decreased to 20 years. At the present time, there is no evidence that ^{131}I induces thyroid carcinoma in adults, but a sufficient time period has not yet elapsed to evaluate the results of ^{131}I therapy on the thyroid in children and adolescents.

The only absolute contraindication to the treatment of hyperthyroidism with ^{131}I is pregnancy. Iodine crosses the placental barrier and localizes in the fetal thyroid, particularly after the first trimester, necessitating another form of treatment.

Thyroid Carcinoma

Cancer of the thyroid is successfully treated with ^{131}I involving doses of several hundred millicuries. A dose of 100 mCi of ^{131}I delivers

50 to 100 rads to the hemopoietic tissues resulting in severe depression of circulating blood cells, particularly white cells. In fact, the limiting factor in the treatment of this disease with ^{131}I is the depression of blood cell counts due to bone marrow depression. Because of the high doses delivered to the bone marrow, an increased incidence of leukemia is certainly probable. Studies have shown a much higher incidence of myeloid leukemia in these individuals which is directly attributable to radiation. However, the risk of leukemia induction following this treatment method is justified in view of the seriousness of the primary disease and the benefits to the patient of ^{131}I therapy.

Polycythemia Vera

The treatment of polycythemia vera (a malignant disease of red blood cells resulting in an increased number of these cells) with radioactive phosphorus (^{32}P) again brings up the question of leukemia. A dose of 4 mCi of ^{32}P delivers 100 rads to the bone marrow, liver and spleen and 10 rads to the total body. Although the high dose to hemopoietic tissues is certainly capable of inducing leukemia, individuals with polycythemia vera appear to have a higher spontaneous incidence of leukemia clouding the picture of radiation effects. The role of radiation in the induction of leukemia in polycythemia vera patients must be evaluated by a controlled study of a large population.

Conclusions

Nuclear medicine is probably the most rapidly growing medical specialty today. Statistics indicate that physicians are referring more patients to nuclear medicine and for more procedures. In some cases diagnostic radiopharmaceutical procedures offer unique information that cannot be gained through other technics. In addition, information is sometimes obtained more rapidly from these procedures than from standard x-ray procedures. Although few diagnostic x-ray studies can be replaced by radiopharmaceutical studies, nuclear medicine is a valuable adjunctive diagnostic aid to all physicians.

No adverse effects have been observed as a result of nuclear medicine; however, the health impact of this specialty remains to be evaluated. In this regard nuclear medicine is in the unique position of influencing concepts and practices of dose reduction that were not possible in the juvenile stages of diagnostic radiology. Establishment of quality control programs assumes major importance in this regard, particularly in view of the federal government's increased regulation and supervision of all medical uses of ionizing radiation. With the rapid expansion, increased complexity and dynamic status of this specialty, "growing pains" are inevitable.

REFERENCES

1. Aboul-Khair, S. A., *et al.*: Structural and functional development of the human foetal thyroid, Clin. Sci. 31:415, 1966.
2. *Continuing Education Lecture, 1971, in Review of Nuclear Medicine,* Vol. 1, Southeastern Chapter of the Society of Nuclear Medicine.
3. *Effects on Populations of Exposure to Low Levels of Ionizing Radiation,* Report of the Advisory Committee on the Biological Effects of Ionizing Radiation, Division of Medical Sciences, National Academy of Sciences/National Research Council, November 1972.
4. *Estimates of Ionizing Radiation Doses in the U.S., 1960–2000,* Environmental Protection Agency, ORP-CSD 72-1, 1972.
5. Gaulden, M. E.: Possible effects of diagnostic x-rays on the human embryo and fetus, J. Arkansas Med. Soc. 70:424, 1974.
6. *Health Physics in the Healing Arts,* Proceedings of the Seventh Midyear Topical Symposium, San Juan, Puerto Rico, December 11–14, 1972, DHEW Publication (FDA) 73-8029, March 1973.
7. Henry, H. F.: *Fundamentals of Radiation Protection* (New York: John Wiley & Sons, 1969).
8. Koch, E. I.: Health legislation. J. Nucl. Med. Tech. 2:20, 1974.
9. *Medical Radiation Information for Litigation,* Proceedings of a Conference, Baylor University College of Medicine, Houston, Texas, November, 21–22, 1968, DMRE 69-3 (Washington, D.C.: U.S. Government Printing Office, 1969).
10. *Medical Radionuclides: Radiation Dose Effects,* Proceedings of a Symposium, Oak Ridge Associated Universities, Oak Ridge, Tennessee, 1969, AEC Symposium Series 20 (Washington, D.C.: U.S. Government Printing Office, 1970).
11. Neil, C. M.: The question of radiation exposure to the hand from handling ^{99m}Tc, J. Nucl, Med. 10:732, 1969.
12. *PDR for Radiology & Nuclear Medicine,* Physician's Desk Reference (Oradell, New Jersey: Medical Economics Company, 1972).
13. *Protection of the Patient in Radionuclide Investigations,* International Commission on Radiological Protection (ICRP), Publication 17 (New York: Pergamon Press, 1971).
14. *Reduction of Radiation Exposure in Nuclear Medicine,* Proceedings of Symposium, August 1967, Environmental Health Series, USDHEW (Washington, D.C.: U.S. Government Printing Office, 1968).
15. Quimby, E. H.: *Safe Handling of Radioactive Isotopes in Nuclear Medicine* (New York: Macmillan, 1960).
16. Smith, E. M.: Internal dose calculations for ^{99m}Tc, J. Nucl. Med. 6:231, 1965.
17. Snyder, W. S.: *Report of the Task Group on Reference Man for Purposes of Radiation Protection* (New York: Pergamon Press, in press).
18. *Survey of the Use of Radionuclides in Medicine,* Stanford Research Institute and USDHEW, Bureau of Radiological Health, STM DMRE 70-1 (Washington, D.C.: U.S. Government Printing Office, 1970).

10/Clinical Radiobiology III: Therapeutic Radiology

Cancer is a leading cause of death in the United States today, second only to heart disease; of every six deaths in the population, one is from cancer. The American Cancer Society has estimated that a total of 358,000 people died from cancer in 1974—960 persons a day, or approximately one person died from cancer every 2 minutes. In this decade there will be an estimated 3.5 million cancer deaths, 6.5 million new cancer cases and 10 million persons under medical treatment for this disease.

Unfortunately, the word "cancer" means inevitable death to many people; however, many individuals have been cured of this disease. Of every six diagnosed cancer patients, two are cured and four die; one of these four deaths could possibly be averted if the disease had been detected and treated earlier. In 1974 alone, 222,000 people were cured of cancer; another 111,000 could have been cured by earlier and better treatment. Today there is a total of 1.5 million Americans living who have been cured of cancer.

Ionizing radiation has long been used as a treatment for cancer, dating back to the discovery of x-rays and their lethal effects on tissues. For many years radiation was regarded as a second treatment choice to be used only when surgery was impossible or had failed. Today approximately 50% of all cancer patients receive radiation for either curative or palliative (relief of symptom) purposes. Over the past two decades radiation therapy has made a great contribution to the improved 5-year survival of many cancer patients, e.g., those with Hodgkin's disease (6% survival rate in 1950; 78% in 1970) and seminoma (52% survival rate in 1950; 94% in 1970), while other diseases, e.g., breast and lung cancer have not reflected these improvements.

Radiation may be used alone in the treatment of some malignant diseases, e.g., cervical carcinoma and early stage Hodgkin's disease, or adjunctively (i.e., in combination) with another form of treatment such as surgery or chemotherapy for diseases such as breast carcinoma. At the present time radiation therapy is a primary treatment modality and plays a significant and vital role in the management of patients with cancer.

Goal of Radiotherapy

A malignant tumor originates and grows in normal tissue, progressively invading and replacing the normal cells with tumor cells and eventually metastasizing (spreading to distant sites) to other normal tissues. The treatment of this tumor with radiation involves delivery of high doses over a period of time to the localized, tumor-bearing area of the body. Because only this small localized area is irradiated (termed the treatment volume), and because the total dose is given in small daily doses over a period of time, total doses on the order of 4000 to 6000 rads and greater can be administered without inducing any of the radiation syndromes or death. These high doses given in this manner are necessary to accomplish the goal of radiotherapy—sterilization of the tumor with minimal destruction of the surrounding normal, noncancerous tissue, ultimately resulting in tumor eradication with sufficient normal tissue remaining to ensure viability and function.

Although appearing at first glance to be an easy task, this is not necessarily so. The interaction of radiation in matter, including cells, tissues and organs, is a nonspecific, random process, with no specificity for tumor cells. The treatment volume encompasses not only the tumor but also surrounding normal tissue. In addition, radiation administered from sources outside the body interacts with and is absorbed by normal tissues in its path en route to the tumor. Not all the radiation is absorbed by the tumor; some continues through the body, eventually exiting at the skin surface, irradiating these normal tissues as well. Because the probability of radiation interacting with tumor cells is the same as the probability of interaction occurring in normal cells, damage is produced in normal tissues in the treatment area as well as in the tumor.

All normal tissues have a limit to the amount of radiation they can receive and still remain functional; this is defined as *radiation tolerance*. For this reason, the amount of radiation used to treat a specific malignant tumor is limited by the tolerance of the surrounding normal tissue, *not* by the tumor. The goal, then, of radiotherapy, although dependent on cellular and tissue response in both the tumor *and* the normal tissue, is ultimately dependent on normal tissue response if complications are to be avoided (which may result in a cure no better, and possibly worse, than the disease).

The differences that exist between the normal tissue and a tumor originating from it are not of sufficient magnitude to ease the task of radiotherapy. Rather, technics must be used that take major advantage of these relatively minor differences between normal and cancerous tissues—such as dose fractionation schedules and the LET of the radiation. Radiobiology has contributed to an understanding of clinically proved technics in radiotherapy as well as to changing and improving these technics. In many ways radiation therapy can be considered "applied radiobiology." Radiobiology has made and continues to make contributions to radiotherapy, but this area of medicine still remains largely empirical with treatment technics primarily determined through clinical experience.

General Tumor Characteristics

Malignant tumors, like normal tissues, are composed of a parenchymal compartment, i.e., the tumor cells and a stromal compartment consisting of normal blood vessels, lymphatic vessels and connective and nerve tissue. In general the parenchymal compartment consists of four subpopulations of tumor cells differing in viability and proliferative status:

Group 1. Viable and actively proliferating (i.e., in the cell cycle).
Group 2. Viable but not proliferating (i.e., not in the cell cycle).
Group 3. Nonviable but intact morphologically.
Group 4. Nonviable and not intact morphologically.

The proportion of tumor cells in each of these subpopulations varies with the size and type of tumor.

GROUP 1.—Observations of both experimental and animal tumors report a variation of 30 to 80% in the population of cells in this group. It is readily apparent that this tumor cell population, termed the *growth fraction*, is responsible for tumor growth and is of clinical concern.

GROUP 2.—Not so readily apparent is the clinical importance of the second population which, although not actively dividing, are viable and capable of proliferating. Initiation of division in these cells, ultimately resulting in tumor recurrence, may be dependent on tumor vasculature and oxygen supply; this topic will be discussed in a later section.

GROUPS 3 AND 4.—Because these are nonviable cells, they cannot produce a recurrence; however, cells in Group 3 may be mistaken as viable due to their intact appearance—an important factor when reassessing radiation effect.

Tumor Growth

Tumor growth, measured in terms of doubling time (the time in which the tumor doubles in volume), is dependent on three factors:

1. The rate of division of the proliferating cells.
2. The growth fraction, i.e., the fraction of the total tumor cell population that is dividing.

3. The degree of cell loss from the tumor.

DIVISION RATE AND GROWTH FRACTION.—The division rate of malignant cells is shorter than that of the normal cells of origin for some tumors. Measured in both animal and human tumors, this observation appears to imply a correspondingly short *tumor* volume doubling time, particularly when compared to normal tissues. Measured tumor doubling times, however, are much longer than expected, ranging from 40 to 100 days. Although a small growth fraction may partially account for these apparently long doubling times, this factor alone cannot explain the discrepancy.

CELL LOSS.—A third important factor is cell loss. Cells are lost from tumors in a number of ways including metastases and cell death. Studies indicate a great variation in cell loss among different types of tumors (0 to 90%), with carcinomas exhibiting a greater cell loss than sarcomas. This observation may be a reflection of the growth characteristics of the normal tissue of origin: carcinomas arise from dividing, constantly renewing, epithelial tissue; sarcomas from nonrenewing cell systems such as connective and soft tissue.

The rate of tumor growth is a balance between these three factors, i.e., cell cycle time, growth fraction and cell loss. These three factors in turn depend largely on vascularization and its influence on supply of oxygen and other nutrients.

Role of Oxygen in Tumor Growth

Malignant neoplasms are more complex tissues than originally believed, appearing to have some internal regulatory growth mechanisms. Unlike normal tissue, however, their unorganized growth pattern results in either an outgrowth of vascular supply or compression of vessels within the tumor. Both these situations produce the same result, a decrease of available oxygen to tumor cells.

A distinct architectural pattern consisting of a central region of necrosis surrounded by a rim of viable cells, termed a cancer cord, was observed in a human bronchial carcinoma by Thomlinson and Gray[45] (Fig. 10-1). Other tumors do not exhibit this characteristic pattern but contain many foci of necrosis. Further studies of the location of these distinctly different morphologic areas revealed them to be related to the size and growth of the tumor. Small tumors with a radius less than $100\,\mu$ did not exhibit a necrotic area, indicating sufficient vascular and therefore oxygen supply. As a tumor grew and its radius exceeded $160\,\mu$, necrotic areas developed surrounded by a rim of viable cells between 100 and 180 μ in thickness. Continued tumor growth resulted in an increase in the size of the necrotic area, but the thickness of the rim of viable cells remained constant (Fig. 10-2). Actual mea-

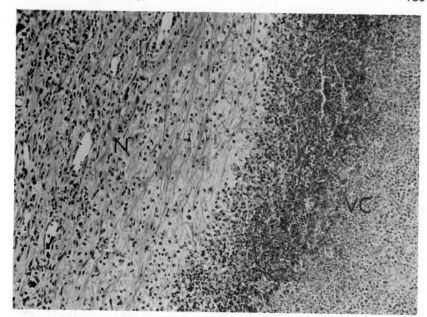

Fig. 10-1.—Photomicrograph of a Walker 256 tumor grown in rats exhibiting a central region of necrosis *(N)* surrounded by a rim of viable cells *(VC)*. The cells visible in the necrotic area are inflammatory cells such as lymphocytes.

surement of oxygen tensions in these two areas demonstrated a much lower oxygen tension in the necrotic than in the viable tumor area, implying either an insufficient vasculature or oxygen diffusion. Apparently oxygen was not only a necessary requirement for life of the tumor cell but also was a determining factor in its ability to proliferate. Indeed, calculation of the diffusion distance of oxygen in tissues by these two investigators revealed a decreasing oxygen concentration with increasing distance from a vessel, approaching zero at approximately 150–200 μ. The close correlation of this calculation with the observed constant width of the rim of viable tumor cells (100–180 μ)

Fig. 10-2.—Diagram of tumor growth illustrating constant width of rim of viable cells (100–180 μ) but increasing size of necrotic area.

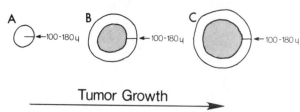

Tumor Growth

supported the experimental supposition that oxygen was a critical, determining factor in tumor growth.

In actuality, a clear-cut border does not exist between well and poorly oxygenated tumor areas; instead, there is a gradient of oxygen tension within a tumor, the highest appearing next to vessels and gradually decreasing with increasing distance from the vessel. Tumors, then, can be considered as consisting of various compartments dependent on oxygen supply. Tumor cells greater than 200 μ from a vessel have no oxygen (are *anoxic*) and will not only be unable to proliferate but also will die, forming a necrotic area. Cells closest to a vessel are well oxygenated and may be actively dividing; these cells comprise the growth fraction. Between these extremes, a gradual decrease of oxygen tension produces increasing degrees of hypoxia; cells in this area, although not dividing, may be viable and capable of dividing (Fig. 10-3). This hypoxic group of cells may comprise 15% of the total number of viable cells.

The vasculature plays a vital role in both tumor growth and the ability of radiation to eradicate the tumor. Cells 200 μ from a blood vessel, nonviable and unable to divide, contribute to a decreased growth fraction; these cells do not pose a problem in radiotherapy. Those cells closest to a vessel also are not a major problem; they have a normal oxygen supply and are relatively sensitive to radiation. However, those cells that are hypoxic but viable may pose the biggest problem because they are relatively resistant to radiation compared to well-oxygenated cells. These cells ultimately may be responsible for regrowth of the tumor.

Fig. 10-3.—Schematic representation of the radius of a tumor illustrating hypothetical tumor compartments related to distance from a vessel and oxygen tension. *Dot* marks the center of this diagrammatic tumor.

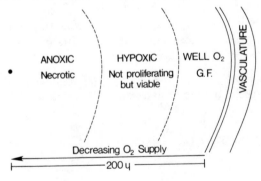

Radiosensitivity and Radiocurability

Tumors, like normal tissues, vary in their sensitivity to radiation; certain types of tumors are more sensitive than others. In general tumor cells reflect the sensitivity of the cell of origin, i.e., those derived from radiosensitive cells are radiosensitive and vice versa. Although the inherent radiosensitivity of a tumor has clinical implications, of primary concern is the probability of curing a tumor with radiation. For this reason a better clinical term indicating both inherent radiosensitivity and the ultimate goal of tumor control is *radiocurability*.

Radiosensitivity and radiocurability, although closely related, are not always synonymous. Radiosensitivity is a function of cell characteristics, i.e., Bergonié and Tribondeau's law. Radiocurability is dependent on many factors, some of which are tumor cell related (radiosensitivity), others which are related to the tumor tissue (rate of metastases and blood supply) and still others related to the patient (age and general condition). In addition, and of primary importance in the clinical situation, is that the term "radiocurability" takes into account the normal tissue in terms of cellular radiosensitivity and radiation response and tolerance. Although there are exceptions, in general, radiosensitive tumors are usually radiocurable while radioresistant tumors usually are not.

Dose Fractionation

Radiation therapy treatment schedules deliver a high total dose to the treatment area by dividing the dose into daily fractions administered over a period of time. That is, the dose is said to be fractionated and the time over which it is given is protracted. The total dose, the size of the fractions, the number of fractions and the time over which the treatment is given are determined by both the tumor type treated and on the tolerance of the surrounding normal tissue. Some tumors, which can be considered radiosensitive, require a higher total dose for cure than do others, which can be considered radioresistant.

Dose fractionation technics originated in 1927 with the observation that cells in the testes were sterilized while producing less damage to the overlying skin if the total dose was divided into fractions than if given in a single exposure. Considering tumor cells comparable to the germinated cells of the testes and the surrounding normal tissue more closely representing the skin, this technic was applied to the treatment of cancer with radiation.

Biologically, fractionated doses are less efficient in causing cell death than are single doses; a higher total dose is necessary to produce the same degree of biologic damage when the dose is fractionated than when it is given as a single dose. Although fractionated doses appear to be advantageous to tumor growth, years of experience have shown them to be more effective in tumor eradication, with a minimum of normal tissue damage, than the same total dose administered singly.

The biologic effect on the tumor and normal tissue of dividing a total dose of radiation into fractions is dependent on an interplay of four factors, termed the "4R's" of radiotherapy: repair, repopulation, reassortment and reoxygenation.

Repopulation

At some time during a multifraction treatment, cells in both the tumor and the normal tissue divide, thereby repopulating these tissues. The reasons for this phenomenon are as yet undefined in both tissues; however, improvement in nutritional status and oxygen may account for tumor repopulation, while normal tissue repopulation may be due to a "feedback" controlling mechanism.

Occurring in both the tumor and in normal tissue, tumor repopulation and regrowth are undesirable, possibly causing a radiotherapy failure, i.e., tumor recurrence. On the other hand, repopulation of the normal tissue is a desirable and necessary factor in the prevention of late complications resulting from exceeding the normal tissue tolerance. Although these two situations appear contradictory, clinical experience indicates that fractionated doses take advantage of repopulation in the normal tissue but not in the tumor, resulting in "sparing" of the former while still eradicating the latter.

Reassortment

Radiation causes a delay in the progression of cells through the cell cycle, subsequently producing reassortment and synchronous progression of cells in their life cycle (Chapter 4). This effect, occurring both in the tumor and in normal tissue, influences the radiosensitivity of the population because radiosensitivity is a function of position of the cell in the cell cycle.

The question that can be asked is whether current fractionation schedules take advantage of this phenomenon by administering succeeding fractions when a large number of tumor cells are in the most sensitive phase. However, the cell cycle position of the normal cells is also of importance. The administration of radiation when tumor cells are in a radiosensitive phase and the normal cells are in a radioresistant phase would be advantageous in terms of tumor cell killing and "sparing" of normal tissue. The opposite situation would be highly undesirable. Unfortunately, it is not feasible to clinically determine the cell cycle position of either population, but it is highly probable that the effects of dose fractionation on reassortment are small and may be accounted for by present treatment schedules, i.e., any radiation-induced synchrony may be overcome in the time interval between fractions.

At the present time many questions remain to be answered concerning reassortment and, although contributions to radiotherapy may be made from further investigation of dose fractionation effects on this phenomenon, it seems to hold the least immediate promise for improvements in radiotherapy. The rationale for the use of synchronizing drugs (e.g., methotrexate and hydroxyurea) in combination with radiotherapy was based on this phenomenon.

Repair

A third factor contributing to the biologic inefficiency of fractionated doses—repair of intracellular sublethal damage (Chapter 4)—occurs within a few hours postirradiation in both normal and malignant cells.

Treatments in radiotherapy are usually administered at 24- or sometimes 48- or 72-hour intervals; sublethal damage is repaired daily and surviving tumor and normal cells essentially respond as nonirradiated cells to the next dose fraction. Occurring in both the tumor and normal tissue, fractionated doses apparently take greater advantage of repair processes in the normal tissue than in the tumor, thus partially accounting for the "sparing" effect on normal tissues of multifraction treatment.

Of particular importance to radiotherapy is the oxygen dependence of repair, i.e., well-oxygenated cells are capable of repair but repair appears inhibited in hypoxic cells. This factor does not play a role in the repair ability of well-oxygenated normal cells but tumors, with their variations in oxygen concentration, are affected.

Well-oxygenated tumor cells, while able to repair sublethal damage, are radiosensitive; a large proportion of this tumor cell population is destroyed daily by fractionated doses. Tumor cells that are very hypoxic are also damaged by fractionated doses because the low oxygen tension inhibits repair. The population of malignant cells that poses a problem is that hypoxic population which has a sufficient amount of available oxygen to repair sublethal damage but has a tension low enough to confer a certain degree of resistance on the cells. This borderline cell population may repopulate the tumor, constituting a radiotherapy failure.

Reoxygenation

The first 3R's apply to both normal and tumor tissues; only the fourth, reoxygenation, applies solely to tumors. Normal tissues are well oxygenated; in contrast, tumors exhibit a gradient of oxygen tension ranging from normal to hypoxic to anoxic. Tumors, then, have an overall lower oxygen tension than normal tissues.

Because radiosensitivity is a function of oxygen tension, various tumor areas are resistant to radiation by a factor as high as 3 (the OER for mammalian cells exposed to acute doses of x- and γ-rays equals 3— Chapter 4). Although the OER is reduced when doses are fractionated, those borderline tumor cells with sufficient oxygen supply to repair sublethal damage and proliferate but still retain some resistance to radiation because of the low oxygen concentration could be the determining factor in the success of radiotherapy.

FATE OF HYPOXIC CELLS DURING RADIOTHERAPY.—For the reasons given above, many models have been postulated concerning the behavior of hypoxic cells during a fractionated treatment schedule. Early models assumed that these cells remained hypoxic throughout the treatment course, requiring a dose for sterilization that exceeded the normal tissue tolerance. Not only was this unacceptable, but doses such as those used in clinical radiotherapy could not have possibly cured a tumor with hypoxic cells—a fact refuted daily in the clinic. Indeed, further study and observation of animal tumors treated with fractionated doses revealed errors in these early models.

Initial dose fractions damage tumor cells in the well-vascularized, well-oxygenated radiosensitive tumor compartment, causing a depopulation in this compartment. Cells in the hypoxic compartment, relatively resistant, are depopulated to a lesser extent producing a significant increase in the proportion of hypoxic cells immediately postirradiation. Within 24 hours, a proportion of the hypoxic tumor cells become reoxygenated and the tumor establishes its original structure of well-oxygenated, hypoxic and anoxic compartments. This change occurring after each dose fraction renders a proportion of hypoxic, radioresistant tumor cells oxygenated and radiosensitive.

Multifraction treatments, the order of the day in radiotherapy, appear to efficiently deal with hypoxic cells and, if human tumors reoxygenate as rapidly as do animal tumors, may allow for daily tumor reoxygenation. This is substantiated by clinical observation of cures in tumors with suspected hypoxic cell populations. This same reasoning can be applied to recurrence of a tumor. Present fractionation schedules may not permit reoxygenation between daily fractions, allowing borderline hypoxic tumor cells to repopulate and cause a recurrence— radiotherapy failure. Alteration of the fractionation schedule theoretically could permit greater reoxygenation and greater radiosensitivity of this population. Hypoxic tumor cells may be one of the most important areas for research, providing the greatest potential for improvements in radiotherapy.

Methods for Overcoming Oxygen Effect

Since reoxygenation is the only factor applying to tumors alone, it is reasonable to attempt to increase oxygen in the tumor, thereby increasing tumor radiosensitivity. Although an increase in tumor oxygen supply also will increase oxygen to the normal tissue, the sensitivity of normal tissue will not be greatly enhanced as it is already well oxygenated and therefore radiosensitive (Fig. 10-4).

The oxygen effect can be overcome by a variety of methods in the laboratory; however, none have resulted in dramatic clinical improvements.

HYPERBARIC OXYGEN.—This method involves placing the patient in a chamber where he or she breathes pure oxygen at 3 atmospheres of pressure. This procedure is time-consuming for two reasons: the patient must be equilibrated in the chamber for 45 minutes prior to and posttreatment daily, and patient positioning is more difficult because he or she must be left in the chamber during treatment. In addition, the highly inflammable and explosive environment creates problems of safety. If this method significantly increased clinical results, it would be worth all efforts; however, this has not occurred. A possible explanation is that oxygen did not reach all hypoxic tumor areas because of its limited diffusion distance (150–$200\,\mu$) from a capillary.

CARBOGEN.—Another method of increasing oxygen concentration in the tumor is breathing carbogen, which is 95% oxygen and 5% carbon dioxide. Carbon dioxide, a vasodilator, assists in increasing the available oxygen to the tumor. In this procedure the patient breathes oxygen through a mask prior to treatment for a given period of time,

Fig. 10-4.—Increasing oxygen tension and its relationship to the relative sensitivity of both normal *(N)* and tumor *(T)* tissue. An increase in oxygen tension significantly increases the sensitivity of the poorly oxygenated tumor *(T¹)*; on the other hand, the sensitivity of normal tissues, because they are well oxygenated, are not as greatly affected *(N¹)*.

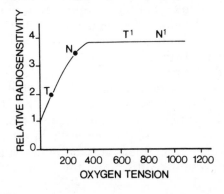

therefore avoiding equilibration, patient setup and safety problems. The effectiveness of carbogen is being clinically tested in the treatment of diseases of the head and neck.

HYPOXIC CELL SENSITIZERS.—Drugs that selectively sensitize hypoxic cells, particularly the nitrofurans such as NDPP (*p*-nitro, 3-dimethylaminopropiophenone hydrochloride) may have potential for radiotherapy. Indeed, if in vitro and in vivo laboratory results with these drugs are any indication, they may be a very exciting development in radiation therapy in the near future. Those drugs which most effectively sensitize hypoxic cells are oxygen mimics; as such, they show little or no oxygen dependence. In addition, these drugs are metabolized slowly and are not limited by a short diffusion distance. In fact, experimentally, they have been shown to diffuse through a nonvascularized mass. Another interesting observation with NDPP in particular is that its sensitizing efficiency appears unrelated to the time interval of injection and irradiation up to 30 minutes, thus overcoming another problem of using oxygen. If further investigation of these drugs supports the already promising experimental in vivo and in vitro results, they will most certainly be tested in the clinic.

LET.—The use of high LET radiations that are less dependent on oxygen for cell killing is a fourth method of overcoming hypoxic cells in radiotherapy. In addition, cell killing by high LET radiations is less dependent on fractionation patterns than are x- or γ-rays (due to small shoulder, less wasted radiation between fractions). For this reason negative pi mesons (*pions*) and fast neutrons are being actively investigated for use in radiotherapy.

The OER for neutrons is 1.7, a significant advantage over the 3.0 OER of x- and γ-rays. Additionally, depth doses from neutron generators are in the range of photons from 250 kvp units to ^{60}Co units, depending on the neutron generator. Though presently used in the treatment of cancer, neutrons are still considered an investigational modality. The prohibitive cost of the equipment presently precludes its use in many hospitals with radiation therapy capabilities, and it is restricted to large cancer centers where patients are referred who will most benefit by this modality of treatment. This trend will probably remain the same in the future.

Pions hold much of the same attraction for radiotherapy as neutrons, possibly more. The OER for pions may be less than the 1.7 of neutrons and also permit better localization of radiation in the tumor with a minimal dose to the surrounding normal tissue, an added attraction to their use. Like neutrons, pions are also an investigational treatment modality and presently suffer from the same high equipment costs.

FRACTIONATION SCHEDULES.—Although standard fractionation schedules (200 rads/day, 5 days/week) apparently allow for reoxygenation of tumor cells, longer time intervals between fractions possibly may achieve greater reoxygenation of the tumor. This factor, in combination with the hoped for increase in repopulation of normal tissue during this interval, is the basis of split course treatments, which will be discussed in another section.

RADIOPROTECTIVE DRUGS.—Although they are not compounds that overcome the oxygen effect in the manner previously discussed, some radioprotective drugs may be potentially useful and feasible in radiotherapy. The mechanism of action of these drugs, rather than sensitizing hypoxic tumor areas, apparently protect well-oxygenated tissues, essentially reducing the damage in the normal tissue while not affecting damage in the tumor. In a sense, then, these drugs do overcome the oxygen effect, but in the opposite manner.

For many years researchers were hampered by the toxic levels required for protection. This problem has been apparently overcome by some drugs, particularly the thiophosphates, an example of which is WR-2721. Renewed interest in the potential use of radioprotective compounds in the radiation therapy clinic has been stimulated by recent laboratory investigations with the thiophosphate compounds.

The Concept of Tolerance

The importance of preserving the integrity of normal tissue was recognized early in radiation therapy, leading to dose fractionation technics. Many different fractionation and protraction schedules and total doses, often referred to as time-dose relationships, were used in these early years to treat malignant disease. In an attempt to correlate these various time-dose relationships with clinical results, Strandqvist[40] reviewed 280 cases of carcinoma of the skin and lip followed for 5 years. Plotting total dose and overall treatment time on a double-log plot, he developed a series of lines, termed isoeffect curves, relating the treatment schedule to clinical results including cure of the disease, occurrence of severe late complications (skin necrosis) and early minor complications such as skin erythema and moist and dry desquamation (Fig. 10-5). Strandqvist observed that some time-dose relationships, although curing the disease, also produced severe late complications while others resulted in no severe late damage but the tumor recurred or persisted (Fig. 10-6). The result of this and other determinations of isoeffect curves, was the establishment of treatment schedules that did not exceed the *tolerance* of various normal tissues but still afforded a high probability of tumor control.

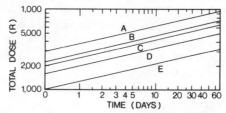

Fig. 10-5.—Isoeffect curves drawn from Strandqvist's data relating various treatment schedules to clinical results: *A,* skin necrosis; *B,* cure of skin cancer; *C,* moist desquamation; *D,* dry desquamation; *E,* skin erythema. (From Strandqvist, M.: Studien über die kumulative Wirkung der Röntgenstrahlen bei Fraktionierung, Acta Radiol. [Suppl.] 55:1–300, 1944. Courtesy of Acta Radiologica.)

From these studies it appeared that the number and size of the fractions are more important in determining tolerance than the overall treatment time.

Tolerance can be defined as the total dose at which additional radiation will significantly increase the probability of occurrence of severe normal tissue reactions, i.e., it defines what dose organs will tolerate. Tolerance is a clinical concept, accounting for inherent radiosensitivity of the parenchyma and stroma, repair capabilities of both, preservation of functional integrity and the importance of the organ to

Fig. 10-6.—Strandqvist data plotted: the area between the two lines represents those time-dose relationships which gave a high probability of cure with a low probability of reactions. Cases above the line resulted in cures with late reactions, while the tumor persisted with no late reactions in those below the line. (From Strandqvist, M.: Studien über die kumulative Wirkung der Röntgenstrahlen bei Fraktionierung, Acta Radiol. [Suppl.] 55:1–300, 1944. Courtesy of Acta Radiologica.)

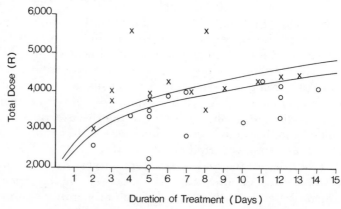

life. A better term than radiosensitivity in the clinical situation, tolerance indicates the variation in response of different organs to the same dose or range of doses. When tolerance is exceeded, the patient exhibits objective and subjective signs and symptoms of damage in the organ.

To better express tolerance, the concept of "tolerance dose" has been put forth. A modification of the $LD_{50/30}$ concept expressing lethality from total body radiation, the tolerance dose predicts the minimum and maximum injurious rates produced in different organs by various time-dose relationships. Because tolerance must be defined in terms of time and because survival from cancer is usually based on 5 years, the tolerance dose is defined as the total dose delivered by a standard fractionation schedule that causes a minimum (5%) or maximum (50%) complication rate in 5 years. The tolerance dose therefore is expressed either as TD 5/5 (that dose which, when given to a population of patients, results in the minimum—5%—severe complications rate within 5 years posttreatment) or TD 50/5 (that dose which, when given to a population of patients, results in the maximum—50%—severe complications rate in 5 years).

Tolerance dose is a clinically practical concept, allowing the therapist to predict with some certainty the risk of inducing severe complications in patients treated using a specific time-dose relationship for a specific malignant disease. Most radiotherapists are willing to exceed the minimum of 5% complications, accepting this morbidity rate.

Tolerance doses of various organs classified as low, moderate, high and very high are presented in Table 10-1. In some cases organs with low tolerance, e.g., gonads and bone marrow, are also considered most radiosensitive (Table 3-4). However, other organs such as lung, liver and kidney, considered relatively radioresistant, are only moderately tolerant. This apparent discrepancy is due to two factors: (1) the difference in evaluation time in the two concepts—tolerance doses are evaluated in terms of years; sensitivity was evaluated after 2 months—again reiterating the importance of time in relation to response and tolerance, and (2) the inclusion of other factors besides inherent cellular radiosensitivity in the concept of tolerance.

Table 10-2 lists the minimum and maximum tolerance doses and the type of injury that occurs based on clinical data. Note both the ranges used for expressing the minimum and maximum levels and the incorporation of volume in the table. The doses are based on standard fractionation schedules, i.e., 1000 rads/week, 200 rads/day and 5 days/week treatment.

TABLE 10-1.—CLASSIFICATION OF ORGAN TOLERANCE DOSES

TOTAL DOSE	ORGANS EXHIBITING COMPLICATIONS
Low: 1000–2000 rads	Gonads: ovaries and testes
	Developing organs, breasts, bone and cartilage
	Bone marrow
	Lens
Moderate: 2000–4500 rads	Stomach, intestine, colon (whole or major portion)
	Liver (3500–4500)
	Kidney (>2500)
	Lung (>2500)
	Heart (>4500)
	Thyroid and pituitary glands
	Developing muscle
	Lymph nodes
High: 5000–7000 rads	Epithelial structure, e.g., skin
	Oral cavity, esophagus, rectum, salivary glands, pancreas, bladder
	Mature bone and cartilage
	Organs of special sense, eye, ear
	CNS—brain and spinal cord
	Adrenals
Very high: >7500 rads	Ureters, vagina, urethra
	Mature breasts
	Muscle, blood, bile ducts

Factors Affecting Tolerance

VOLUME.—Because volume affects response, it will, of course, greatly affect the tolerance of an organ to radiation. The volume dependence of response is related to the concept of *integral dose*. Defined in gram-rads, integral dose incorporates the volume, in grams, of tissue irradiated. Per given rad dose, large volumes, including more grams of tissue, have a higher integral dose than small volumes. For this reason large volumes tolerate less radiation than small volumes, i.e., a lower dose must be given when a large volume is irradiated if the acceptable complication rate is not to be exceeded.

FRACTION SIZE.—For many years daily doses of 200 rads were routinely used in radiotherapy. Today, however, larger daily doses are sometimes given over shorter periods of time, e.g., in split course treatment schedules. Although a sufficient amount of data has not been collected, it is feasible that these larger daily doses will result in more damage in normal tissue appearing clinically as a decrease in tolerance. This certainly is accounted for by radiobiologic concepts.

Fractionation schedules requiring daily doses of 400 rads result in much greater cell killing in both the tumor and normal tissue than smaller daily doses (200 rads to HeLa cells in vitro result in a surviving fraction of 0.48, while the surviving fraction following 400 rads

TABLE 10-2.—RADIATION TOLERANCE DOSES AND
COMPLICATIONS*

ORGAN**	INJURY AT 5 YEARS	1–5% TD 5/5	25–50% TD 50/5	VOLUME OR LENGTH
Skin	Ulcer, severe fibrosis	5,500	7,000	100 cm³
Oral mucosa	Ulcer, severe fibrosis	6,000	7,500	50 cm³
Esophagus	Ulcer, stricture	6,000	7,500	75 cm³
Stomach	Ulcer, perforation	4,500	5,000	100 cm³
Colon	Ulcer, stricture	4,500	6,500	100 cm³
Rectum	Ulcer, stricture	5,500	8,000	100 cm³
Liver	Liver failure, ascites	3,500	4,500	Whole
Kidney	Nephrosclerosis	2,300	2,800	Whole
Bladder	Ulcer, contracture	6,000	8,000	Whole
Testes	Permanent sterilization	500–1500	2,000	Whole
Ovary	Permanent sterilization	200–300	625–1,200	Whole
Uterus	Necrosis, perforation	>10,000	>20,000	Whole
Vagina	Ulcer, fistula	9,000	>10,000	5 cm³
Breast (child)	No development	1,000	1,500	5 cm³
Breast (adult)	Atrophy and necrosis	>5,000	>10,000	Whole
Lung	Pneumonitis, fibrosis	4,000	6,000	Lobe
			2,500	Whole
Capillaries	Telangiectasia, sclerosis	5,000–6,000	7,000–10,000	—
Heart	Pericarditis, pancarditis	4,000	>10,000	Whole
Bone (child)	Arrested growth	2,000	3,000	10 cm³
Bone (adult)	Necrosis, fracture	6,000	15,000	10 cm³
Cartilage (child)	Arrested growth	1,000	3,000	Whole
Cartilage (adult)	Necrosis	6,000	10,000†	Whole
CNS	Necrosis	5,000	>6,000	Whole
Spinal cord	Necrosis, transection	5,000	>6,000	5 cm³
Lens	Cataract	500	1,200	Whole
Thyroid	Hypothyroidism	4,500	15,000	Whole
Pituitary	Hypopituitarism	4,500	20,000–30,000	Whole
Muscle (child)	No development	2,000–3,000	4,000–5,000	Whole
Muscle (adult)	Atrophy	>10,000	—	Whole
Bone Marrow	Hypoplasia	200	550	Whole
		2,000	4,000–5,000	Localized
Lymph nodes	Atrophy	4,500	>7,000	—
Lymphatics	Sclerosis	5,000	>8,000	—
Fetus	Death	200	400	—

*From Rubin, P., and Casarett, G. W.: *Clinical Radiation Biology*, Vol. II (Philadelphia: W. B. Saunders, 1968).
**There is no dose-data available for the pancreas, gallbladder or aorta.
†Absorbed dose.

is 0.15). In addition, higher daily doses permit less repair between fractions. Though both of these situations (increased cell killing and decreased repair) are advantageous for tumor control, they may be disadvantageous in the normal tissue, resulting in an unacceptable frequency of late clinical complications. Therefore, the tolerance dose of normal tissues may have to be decreased when treatment schedules employing high daily doses are utilized. This phenomenon already has been observed in the spinal cord. At doses of 200 rads/day, a total dose of 5000 rads to a small volume of cord is generally considered tolerance; however, at 400 rads/day, daily doses given during some split course schedules for treatment of lung cancer, the tolerance dose may be as low as 2500 rads.

Nominal Standard Dose (NSD)

The different time-dose relationships in radiotherapy used to treat various malignant diseases are usually expressed as total tumor dose and number of days over which treatment is given, e.g., 5000 rads in 5 weeks and 6000 rads in 6 weeks. Inherent in these figures is the understanding that these schedules do not exceed the tolerance of normal tissues surrounding the tumor. Because treatment schedules for the same disease may vary among radiotherapists, it has always been desirable to have one figure incorporating all these factors to enable comparisons to be made between different treatment schedules and their biologic effect on the tumor and normal tissue. The concept of *nominal standard dose* (NSD) has been proposed by Ellis[15] as this unit.

Because radiotherapy is limited by normal tissue tolerance, NSD is based on the normal tissue and is calculated from an equation derived from isoeffect curves. The equation proposed,

$$D = (NSD)T^{0.11} N^{0.24}$$

where Dose is total dose, N is the number of fractions and T is the overall time of treatment, is based on the isoeffect curves for both skin tolerance and cure of skin carcinoma; these curves yielded exponents 0.24 and 0.11.

NSD, expressed in *rets* (rad equivalent therapy), is based on normal tissue tolerance, specifically connective tissue tolerance. It has been proposed that one NSD be calculated for a radiotherapy center because connective tissue throughout the body is the same. For many departments, the NSD is 1800 rets. One of the major drawbacks to NSD is it does not incorporate a necessary and vital tolerance parameter—volume—in the equation.

NSD primarily is used to compare two different treatment schedules; however, it also may be used to evaluate split course schedules or to change a treatment plan at any point during treatment.

NSD is not the only answer to the problem of equating different treatment schedules. Although some individuals question the concept, NSD, if used correctly, may be of value in radiotherapy. A sample problem is given for the calculation of NSD:

What is the NSD for a treatment schedule giving a total dose of 6000 radians in 30 fractions?

T = 39 days (time from first day of treatment to the last)
N = 30
D = 6000
$6000 = NSD \times 39^{0.11} \times 30^{0.24}$
NSD = 1773 rets

Tables have been prepared that eliminate the need for these calculations by presenting the combined expression:

$$T^{0.11} \times N^{0.24}$$

for various treatment schedules. Using these tables, it is necessary
only to look up the factor for the number of days and number of fractions and to multiply this by the total dose. More accurate tables, consisting of TDF (time, dose and fractionation) factors that account for
the number of treatment days per week, have been calculated by
Orton [29]. These tables may be used by any radiotherapy center because they are not dependent on a specific NSD.

Treatment Technics

Many and varied treatment technics are used in radiotherapy. In
some cases the variable is dose-rate: in others it is the fractionation
schedule; while in still others it is the number of fields treated at each
session. This section will discuss these various treatment technics and
their biologic rationale.

Dose-Rate

Dose-rates in most radiotherapy centers range from moderate
(greater than 100 rads/min), such as those used in external beam therapy, to low (40 rads/hour), such as those used in interstitial and intracavitary treatment with sealed sources, e.g., ^{226}Ra (radium), ^{137}Cs (ce-
sium), ^{252}Cf (californium) and ^{222}Rn (radon). Mammalian cells exhibit a
dose-rate effect: low dose-rates permit more repair than high dose-
rates reflected as a broader shoulder and more shallow slope of the
survival curve, indicating the biologic ineffectiveness of low dose-
rates (Chapter 4). In spite of this, the use of low dose-rate radiations is
a tried and true technic in the clinical situation, e.g., intracavitary
radium applications in the treatment of cervical cancer.

The clinical use of radium is based on specific physical properties: dose from a radium source falls off rapidly with distance, thereby
delivering a high dose to a localized area, i.e., the tumor, while
sparing the normal tissue. However, ^{226}Ra is a mixed α and γ emitter,
with the γ component (the effects of which are oxygen-dependent) the
primary contributor to dose and therefore biologic effect. Because intracavitary treatment is given toward the end of a treatment course,

when a large proportion of hypoxic cells may exist in the tumor, a less oxygen-dependent radiation could provide better results. For this reason ^{252}Cf, a mixed neutron and γ emitter, is proposed as a replacement for radium.

Dose-rates of 20 rads/hour utilize the neutron component of ^{252}Cf as the primary contributor to dose; however, neutrons require special shielding and pose more hazards and problems in terms of personnel protection than do x- or γ-rays and are more inconvenient to use than radium. In addition, the RBE for neutrons increases with decreasing dose-rates, therefore the stated RBE of 10 may be an underestimate of its true value at lower dose-rates.

At the other end of the dose-rate spectrum is ultrahigh dose-rate radiation (e.g., electrons at 1000 rads/minute), such as those produced by some linear accelerators. Although producing more damage in the tumor, experimental evidence indicates a corresponding increase in normal tissue damage. Further research on ultrahigh dose-rates is necessary to determine their feasibility for use in radiotherapy.

Treatment Schedules

For many years classical fractionation schedules dominated radiation therapy treatment. Recent years have seen a change in this technic to 3 days/week, 4 days/week or split course treatment schedules. Although by no means new and novel ideas, some of these technics have gained acceptance, particularly split course treatments, while others have been abandoned, particularly 3 days/week schedules, because of their associated increased morbidity.

Appropriately named, split course treatment schedules involve administering the total dose in two segments, i.e., courses, separated by a variable rest interval during which treatment is not given. One example of this type of treatment schedule is 4000 rads administered in two courses of 2000 rads over 5 days separated by a rest period of 2 or 3 weeks. Generally, split course schedules deliver lower total doses but higher daily doses (400 rads in the example above) than standard fractionation schedules. Based on the 4R's of radiotherapy, tumor shrinkage should occur during the rest period, allowing for better tumor reoxygenation, while normal tissues should repopulate during this same interval. In general, split course schedules, theoretically, should produce better clinical results, i.e., more tumor cell killing with more sparing of normal tissue, than standard treatment schedules. However, the higher daily doses employed, while advantageous to tumor cell killing, appear to be a disadvantage to normal tissue repopulation, clinically indicated by a decrease in tolerance.

Although split course treatment schedules may be valuable in the treatment of some malignant diseases, standard treatment schedules are still clinically proved and valuable in many others. In fact, the trend to split course schedules, so prevalent a few years ago, appears to be reversing with standard treatment schedules again coming to the fore.

Treatment of All Fields at Every Session

A trend toward treatment of all fields at every treatment session is gaining widespread acceptance. Although more time-consuming than treating one field per session imposing hardships on busy departments and those without rotational units, this technic permits more sparing in the normal tissue.

The importance of treating all fields at each session was recognized by early radiotherapists using 250-kvp x-rays. Because of the poor penetration and therefore minimal skin sparing effect (i.e., normal tissue sparing), they found it more advantageous to treat all fields at each session. Although higher energies (4 mev and ^{60}Co) result in increased depth dose and therefore less normal tissue damage, treatment of all fields at each treatment session still produces more normal tissue sparing than treatment of one field. The basis for this effect can readily be seen from examination of a cell survival curve (Fig. 10-7).

Assuming a 50% depth dose, a daily tumor dose of 200 rads delivered by one field technic will deliver 400 rads to the superficial normal tissues resulting in approximately 0.15 surviving cell fraction

Fig. 10-7.—Survival curve illustrating the radiobiologic rationale for treating all treatment fields at each treatment session. A, 200 rads results in a cell survival fraction of 0.50. B, 400 rads results in a 0.15 surviving fraction, less than ⅓ of the 200 rad dose. See text for further discussion. (From Wilson, C. S. and Hall, E. J.: On the advisability of treating all fields at each radiotherapy session, Radiology 98:419, 1971.)

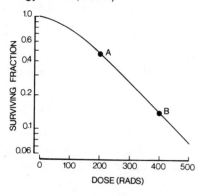

in these tissues. In comparison, if 200 rads is delivered by two fields (100 rads from each field), the tumor still receives a total dose of 200 rads but the surrounding normal tissue would receive only 200 rads resulting in a surviving cell fraction of approximately 0.5.

In actual numbers, if the normal tissues contain 1 million cells, 150,000 would survive treating one field per session, but 500,000 would survive treating all fields per session. After two treatments 75,000 would survive treating one field per session, but 250,000 would survive treating all fields each session. Logically, as treatment continues, appreciably more cells are killed by one field technic than by two. This could mean the difference between either severe late normal tissue reactions or minimal late reactions. For this reason all fields should be treated at each treatment session, particularly when higher daily doses are used, in especially heavy patients, or when the given dose through one field exceeds 400 rads.

The advantages of treating multiple fields at each treatment session are much less with higher energies such as 20 megavolt photons, due to the increased depth dose.

Combined Treatment Modalities

An increased knowledge of cancer, discovery of new forms of treatment and years of clinical experience have shown that a multidisciplinary approach (a combination of treatment modalities) offers a higher probability of cure in certain malignant diseases than one modality of treatment. For this reason, modern cancer management and treatment may involve the radiotherapist, the surgeon, the chemotherapist and—new on the scene—the immunotherapist. Because radiation may be used in conjunction with any one or combination of these modalities, and because increasing numbers of patients are receiving more than one form of therapy, the rationale for combining modalities is important to understanding clinical radiotherapy.

RADIATION PLUS SURGERY.—Radiation is often combined with surgery, either pre-operatively or postoperatively, in the management of some malignant diseases, e.g., lung and breast cancers. Radiation may be used to

1. Eradicate tumor cells outside the surgical margin.
2. Eradicate tumor cells spilled in the surgical field.
3. Diminish the incidence of distant metastases.
4. Shrink large tumor masses prior to surgery, rendering them resectable.

Although opinions have varied throughout the years regarding the use of surgery and radiation and the doses to be used, a combination of modalities remains an accepted method of treating some types of cancer.

RADIATION PLUS CHEMOTHERAPY.—The use of drugs in combination with radiation is becoming an increasingly popular treatment for many malignant diseases. Chemotherapeutic drugs act in one of two ways with radiation:
1. Adjunctive—drugs and radiation kill tumor and normal cells in an additive manner.
2. Potentiate—the combined effect of both is greater than the effect of the sum of each individual treatment.

Radiosensitizing drugs are potentiators; they have no marked effect on tumor cells by themselves but greatly enhance (potentiate) the effects of radiation. Drugs that act in this manner are preferred in combined therapy.

True radiosensitizers that have been clinically tested are 5-BUdR and 5-IUdR; sensitization occurs only if these drugs are incorporated in the cell, more specifically, in DNA. However, both agents are nonselective; as such, they are incorporated in both tumor and normal cells. Primarily due to the inability to localize the drug in the tumor, no advantage is gained in this situation because normal tissue sensitivity increases correspondingly with tumor sensitivity.

Drugs such as methotrexate (MTX) and hydroxyurea (HU), although not potentiators, synchronize the population of cells. It was hoped that use of these drugs would permit timing of radiation in relation to an increased sensitivity of the tumor cell population (i.e., when more cells are in a radiosensitive stage of the cell cycle). However, it is difficult to clinically determine this time and therefore the best time for irradiation.

Bleomycin, a new drug that itself is effective in cell killing, localizes in squamous epithelium and is presently being used in the treatment of squamous carcinoma, melanoma and Hodgkin's disease. Although data are still being accumulated, bleomycin appears to potentiate the effect of radiation. The true test of its usefulness awaits further collection of data both in the laboratory and in the clinic.

RADIATION PLUS IMMUNOTHERAPY.—At this point a word is necessary about immunotherapy. The immune system includes those cells (lymphocytes and other white cells) and organs (spleen, lymph nodes and thymus) that provide natural defense against disease to the body. For many years people have postulated that cancer may be related to abnormal functioning of the immune system; therefore, modification of this system may enable it to function and destroy the tumor. Major research efforts are presently directed toward elucidating the effect of cancer on this system. If a relationship exists and if the immune system can be modified, immunotherapy then will become another modality of cancer treatment and, in all likelihood, will be combined with radiotherapy.

Because the study of the immune system is still in its infancy, knowledge of the response of this system to radiation is minimal. It is well established that radiation causes a depression of immune cells in the circulating blood. In addition, current research indicates that radiation depresses the functional ability of the immune system. Although the relationship of these findings to radiotherapy is as yet undefined, they do not contraindicate the use of radiation as a treatment modality.

The future will probably see a combination of radiation therapy and immunotherapy in the management and treatment of cancer. However, much more information is necessary concerning the immune system, its relationship to cancer and the effects of radiation on the system before these modalities can be combined in the clinic.

Conclusions

The dictum of any medical treatment is
FIRST—DO NO HARM!
Because ionizing radiation is nonselective and because radiation therapy unavoidably exposes normal tissue, this maxim has increased importance in this clinical specialty.

The administration of radiotherapy involves the efforts of many individuals including a radiation therapist, radiation physicist, radiation biologist, therapy technologist, dosimetrist and, in many departments, a nurse. Although each individual functions in a different area of treatment, all are equally important in the proper care of the patient. Each of these individuals is expected to have sufficient knowledge to appreciate both the dangers involved in the treatment of cancer with ionizing radiation and the importance of precautions. Although the radiotherapist assumes ultimate responsibility for the patient, knowledge of radiobiologic concepts applicable to radiotherapy allows each of these individuals a better understanding of the rationale for treating each patient and the implications of the treatment in the disease course.

Because the dose necessary to obtain a cure is usually very close to the tolerance of the surrounding normal tissue, daily accuracy and care are axioms in radiation therapy. The importance of accurate daily positioning of the patient cannot be overemphasized. Although the treatment field includes a small amount of normal tissue, it is not of sufficient magnitude to allow "wandering fields" without producing either a geographical miss, and therefore persistent or recurrent disease, or an unexpected late reaction, caused by exceeding the normal tissue tolerance. All of the thorough pretreatment planning can be negated by careless and improper positioning of the patient under the treatment unit.

Accurate setting of daily doses on the machine is extremely important; a 10% decrease in dose may result in tumor recurrence, while a 10% increase may exceed normal tissue tolerance. Treatment aids such as wedges or bolus always should be included when prescribed; negligence in this area can adversely affect the outcome of the treatment, resulting in patient morbidity. Negligence in accurately following the treatment plan, for example, equal loading of prescribed unequally loaded fields or treatment of only one field when two are indicated also can produce increased patient morbidity.

Knowledge of normal tissue response enhances the position of each individual on the team, allowing the technologist and nurse to monitor all patients daily for unusual and unexpected reactions, noting and reporting these immediately to the physician for care. The physicist and dosimetrist also must have a grasp of this concept if they are to provide the therapist with dosimetry plans that are feasible and reasonable.

Radiobiology provides the basis for radiotherapy, both conceptually and practically in the day-to-day care and treatment of the patient. Radiobiologic investigations are sometimes retrospective in nature, providing an explanation of observed clinical results, as well as prospective, providing the experimental rationale for new clinical technics. Unfortunately, what works in the radiobiology laboratory is not always feasible in the radiotherapy clinic. Further advances in radiation therapy probably will arise both from the physics world, such as the use of neutrons and pions in treatment, and from the radiobiology world (i.e., a greater understanding will be gained of the effects of radiation on various malignant diseases, all of which are different from each other, of the factors affecting normal tissue response and of the manipulation of present technics and implementation of new and improved technics that take greater advantage of the difference in normal and cancerous tissues).

All individuals on the radiotherapy team have important functions to fulfill in the administration of this treatment modality. Thorough knowledge of their respective areas and open communication with all individuals on the team can only enhance the vital role of contemporary radiotherapy in the care and management of the cancer patient, thus making a significant contribution to the number of patients cured of this disease.

REFERENCES

1. Adams, G. E., *et al.*: Electron affinic sensitization, Int. J. Radiat. Biol. 19:575, 1969.

2. Armstrong, D. I.: NSD calculations—A simple graphical method (Correspondence), Br. J. Radiol. 47:363, 1974.
3. Bagshaw, M. A.: Possible role of potentiators in radiation therapy, Am. J. Roentgenol. Radium Ther. Nucl. Med. 85:822, 1961.
4. Barendsen, G. W.: Possibilities for the application of fast neutrons in radiotherapy; Recovery and oxygen enhancement of radiation-induced damage in relation to linear energy transfer, Eur. J. Cancer 2:333, 1966.
5. Belli, J. A., et al.: Radiation response in mammalian tumor cells, I. Repair of sublethal damage in vivo, J. Natl. Cancer Inst. 38:673, 1967.
6. Bewley, D. K.: Radiobiological research with fast neutrons and the implications for radiotherapy, Radiology 86:251, 1966.
7. Bleehen, N. M.: Biological basis of radiotherapy, Br. Med. Bull. 29, 1973.
8. Brown, J. M., and Berry, J.: Effects of x-irradiation on cell population kinetics in a model tumor and normal tissue system; Implications for treatment of human malignancies, Br. J. Radiol. 42:372, 1969.
9. Cancer Facts & Figures, American Cancer Society Report, 1975.
10. Cellular Radiation Biology, Proceedings of the Nineteenth Annual Symposium on Fundamental Cancer Research, Houston, 1964 (Baltimore: Williams & Wilkins, 1965).
11. Conference on Time and Dose Relationships in Radiation Biology as Applied to Radiotherapy, Carmel, California, September 1969, BNL Report 50203-C-57, Biology in Medicine TID 4500 (Upton, New York: Brookhaven National Laboratory, 1969).
12. Denekamp, J.: The cellular proliferation kinetics of animal tumors, Cancer Res. 30:339, 1970.
13. Elkind, M. M., and Sutton-Gilbert, H.: Radiation response of mammalian cells grown in culture, I. Repair of x-ray damage in surviving Chinese hamster cells, Radiat. Res. 13:556-593, 1960.
14. Ellis, F.: Dose, Time, and Fractionation in Radiotherapy, in Ebert, M., and Howard, A. (eds.): Current Topics in Radiation Research (Amsterdam: North Holland Publishing Company, 1968), pp. 359–397.
15. Ellis, F.: Dose, time, and fractionation, a clinical hypothesis, Clin. Radiol. 20:1, 1969.
16. Fowler, J. F.: Radiotherapy, the next decade (abridged), Proc. R. Soc. Lond. [Biol.].
17. Fowler, J. F., and Stern, B. E.: Dose rate effects: Some theoretical and practical considerations, Br. J. Radiol. 33:389, 1960.
18. Fowler, J. F.: Biological Foundations of Radiotherapy, in Turano, L., et al. (eds.): Progress in Radiobiology, Vol. 1, Proceedings of the Eleventh International Congress, Rome, 1965 (Amsterdam: Exerpta Medica Foundation, 1967), pp. 731–737.
19. Fowler, J. F.: Current aspects of radiobiology as applied to radiotherapy, Clin. Radiol. 23:257, 1972.
20. Hall, E. J.: Radiation dose rate; A factor of importance in radiobiology and radiotherapy, Br. J. Radiol. 45:81, 1972.
21. Lindop, P. J.: Radiotherapy, radiobiology; Can radiobiology contribute? Br. J. Radiol. 46:799, 1973.
22. Lajtha, L. G., and Oliver, R.: Some radiobiological considerations in radiotherapy, Br. J. Radiol. 34:252, 1961.
23. Lyskin, A. B., and Mendelsohn, M. L.: Comparison of cell cycle in induced carcinomas and their normal counterparts, Cancer Res. 24:1131, 1964.

24. McCredie, J. A., *et al.*: Rate of tumor growth in mammals, Growth 29:331, 1965.
25. Mendelsohn, M. L.: The growth fraction; A new concept applied to tumors, Science 132:1496, 1960.
26. Deeley, T., and Wood, C.: *Modern Trends in Radiotherapy*, Vol. 1 (London: Butterworths, 1967).
27. Deeley, T.: *Modern Trends in Radiotherapy*, Vol. 2 (London: Butterworths, 1972).
28. Mottram, J. C.: Factors of importance in radiosensitivity of tumors, Br. J. Radiol. 9:539, 1955.
29. Orton, C. G.: Analysis and discussion of the time, dose, fractionation problem, AAPM Q. Bull. 6:173, 1972.
30. Orton, C. G., and Ellis, F.: A simplification in the use of the NSD concept in practical radiotherapy, Br. J. Radiol. 46:529,1973.
31. Phillips, T. L., *et al.*: Radiation protection of tumor and normal tissue by thiophosphate compounds, Cancer 32:528, 1973.
32. Powers, W. E., and Tolmach, L. J.: Demonstration of anoxic component in mouse tumor cell populations by in vivo assay of survival following radiation, Radiology 83:328, 1964.
33. *Proceedings of the 2nd International Symposium on Radiosensitizing and Radioprotective Drugs*, Rome, 1969 (London: Taylor and Francis, 1970).
34. Raju, M. R., and Richmond, C.: Physical and radiobiological aspects of negative pions with reference to radiotherapy, Gann Monograph 9, 1970, pp. 105–121.
35. Rao, G. U. V., and Hazra, T. A.: Time, dose, and fractionation factors in radiotherapy (Correspondence), Br. J. Radiol. 47:432, 1974.
36. Rufsome, S. B., and Berdal, D.: Cell loss in malignant tumors in man, Eur. J. Cancer 3:235, 1967.
37. Sinclair, W. K., and Moreton, R. A.: X-ray sensitivity during the cell generation cycle of cultured Chinese hamster cells, Radiat. Res. 29:450, 1966.
38. Steele, G. G., and Lamerton, L. F.: The growth rate of human tumors, Br. J. Cancer 20:74, 1966.
39. Steele, G. G.: Cell loss as a factor in the growth rate of human tumors, Eur. J. Cancer 3:381, 1967.
40. Strandqvist, M.: Studien über die kumulative Wirkung der Röntgenstrahlen bei Fraktionierung, Acta Radiol. (Suppl.) 55:1, 1944.
41. Suit, H. D., and Maeda, M.: Hyperbaric oxygen and radiobiology of the C_3H mouse mammary carcinoma, J. Natl. Cancer Inst. 39:639, 1967.
42. Suit, H. D., and Urano, M.: Repair of sublethal radiation injury in hypoxic cells of the C_3H mouse mammary carcinoma, Radiat. Res. 37:423, 1969.
43. Tannock, I. F.: The relation between cell proliferation in the vascular system and the transplanted mouse mammary tumor, Br. J. Cancer 22:258, 1968.
44. Terasima, T., and Tolmach, L. J.: Variations in several responses of HeLa cells to x-irradiation during division cycle, Biophys. J. 3:11, 1965.
45. Thomlinson, R. H., and Gray, L. H.: The histological structure of some human lung cancers and the possible implications for radiotherapy, Br. J. Cancer 9:539, 1955.
46. Thomlinson, R. H.: Effect of fractionated irradiation on proportion of anoxic cells in an intact experimental tumor, Br. J. Radiol. 39:158, 1966.

47. Tubiana, M., *et al.:* The application of radiobiologic knowledge and cellular kinetics to radiation therapy, Am. J. Roentgenol. Radium Ther. Nucl. Med. 102:822, 1968.
48. Van Putten, L. M., and Kahlman, L. F.: Oxygenation status of the transplantable tumor during fractionated radiotherapy, J. Natl. Cancer Inst. 40:441, 1968.
49. Yuhas, J. M., and Storer, J. B.: Differential chemoprotection of normal and malignant tissues, J. Natl. Cancer Inst. 42:331, 1969.

Glossary

Aberration: Atypical development or growth.

Accident Y-12: Reactor accident at the Y-12 plant, Oak Ridge, Tennessee, in 1958, in which five persons were exposed to mixed neutron-gamma radiation, receiving doses estimated at 298–461 rems; all incurred radiation injuries but recovered.

Acentric: Without a center; in genetics, denoting a chromosome fragment without a centromere.

Actinomycin D: An antitumor antibiotic that appears to be a radiosensitizer; has been used alone and in combination with radiation for the treatment of cancer.

Adenine: One of the two purines found in both RNA and DNA.

Albumin: Egg white—a well-known protein.

Alopecia: Loss of hair.

Alpha Particle (α-Particle): A positively charged nuclear particle consisting of 2 protons and 2 neutrons with a mass of 4 and a charge of +2; identical to the nucleus of a helium atom; its physical properties render it highly ionizing, and therefore it is considered a high LET radiation.

Amino Acid: Building block of protein or an organic acid linked by peptide bonds to form a protein.

Anabolism: The building up or synthesis of organic compounds.

Anaphase: The stage of mitosis or meiosis in which the chromosomes move from the equatorial plate toward the poles of the cell.

Ancel and Vitemberger, Law of: The inherent susceptibility of any cell to damage by ionizing radiation is the same, but the time of appearance of radiation-induced damage differs among different types of cells, named after the two investigators responsible for this determination.

Anemia: A condition in which the blood is deficient in red blood cells, hemoglobin or total volume; usually manifested by pallor of the skin, lack of energy, shortness of breath and fatigability.

Angiography: Radiography of vessels following the injection of a radiopaque material into an artery, e.g., the carotid.

Ankylosing Spondylitis: Closely related condition to, or variant of, rheumatoid arthritis.

Anomaly: Deviation from the average or norm; anything unusual or irregular or contrary to a general rule.

213

Anophthalmia: Absence of one or both eyes.

Anoxia: Total oxygen deficiency.

Armamentarium: Equipment and methods used in a particular field, often applied to medicine.

Arteriography: Visualization of an artery or arteries by x-rays after injection of a radiopaque contrast medium.

Ascites: Accumulation of serous fluid in the abdomen.

Asynchronous: Not occurring simultaneously.

Atrophy: Indicates decrease in size of cells in an organ, possibly producing a decrease in size of organ.

Basal Layer: The immature, dividing layer of cells at the base of the epidermis.

BEIR: Biological Effects of Ionizing Radiations; advisory committee that published a report on the effects of ionizing radiation on biologic systems.

Bergonié and Tribondeau, Law of: Ionizing radiation is more effective against cells that are actively mitotic, undifferentiated and have a long dividing future; named after the two investigators who determined this effect; considered a byword in radiobiology.

Beta Particle: An electron (β^-) or positron (β^+) ejected from the nucleus of an atom during radioactive decay has mass identical to an electron and a charge of -1 or $+1$, respectively. β-particles from ^{32}P are used in the treatment of polycythemia vera.

Biologic Half-Life: The time taken for one-half of an administered radioactive substance to be lost through the biologic process of elimination.

Bleomycin: An antitumor antibiotic that localizes in squamous epithelium and is presently being used in the treatment of many types of cancer, e.g., squamous carcinoma, melanoma and Hodgkin's disease either alone or in combination with another form of treatment, appears to potentiate the effect of radiation.

Blood Platelet: An irregularly shaped disk, about one-third to one-half the size of an erythrocyte, containing no hemoglobin; necessary for blood clotting.

Bolus: A tissue equivalent material placed on the surface of the body to minimize the effects of an irregular surface; used in external beam radiation therapy treatments to assure more uniform dose distribution.

Bone Marrow: The tissue filling the cavities of bones and having a stroma of reticular fibers and cells; it may be yellow because the meshes of the reticular network are filled with fat or red because the meshes contain the precursors to the mature circulating blood cells.

Brachytherapy: "Short distance" radiotherapy, using sealed sources, intracavitary radium for cervical carcinoma.

Californium 252: A mixed neutron and γ-emitter, sealed sources are used in brachytherapy.

Cancer: A general laymen's term frequently used to indicate any of the various types of malignant neoplasms.

Cancer Cord: Named for a distinct architectural pattern observed in a human bronchial carcinoma, consisting of a central region of necrosis surrounded by a rim of viable cells.

Capillary: Any of the smallest vessels of the blood-vascular system connecting arterioles with venules and forming networks throughout the body.

Carbogen: 95% O_2 and 5% CO_2, breathing of the mixture may increase oxygen concentration in a tumor; CO_2, a vasodilator, assists in increasing the available oxygen to the tumor.

Carbohydrate: An organic compound composed of carbon, hydrogen and oxygen and the primary source of energy to the cell.

Carcinogenesis: The origin or production of all malignant neoplasms.

Carcinoma: A malignant tumor arising from any epithelium in the body; e.g., carcinoma of the cervix, carcinoma of the larynx and carcinoma of the skin.

Cardiac Catheterization: The passage of a catheter (a hollow cylinder of silver, rubber, etc.) into the heart.

Cardiovascular: Relating to the heart, blood vessels or circulation.

Catabolism: The breaking down of organic compounds to provide energy and other requirements for life.

Catalyst: Compounds that increase the rate of chemical reactions.

Cataract: A clouding of the lens of the eye or its capsule, obstructing the passage of light and causing visual impairment; an opacity of the lens of the eye.

Cell: The smallest unit of protoplasm capable of independent existence.

Centrifugation: The process by which particles in suspension in a fluid may be separated; this is done by whirling a vessel containing the fluid about in a circle, the centrifugal force throwing the particles to the peripheral part of the rotated vessel.

Centrioles: Structures of the cell to which the spindle fibers are attached during cell division.

Centromere: A clear region of the chromosome that is necessary to the movement of the chromosome during cell division.

Cervix: Neck or any neck-like structure; especially the *neck* of the uterus, connecting the vagina with the uterine cavity; also used in reference to structures in the neck region of the body; e.g., cervical spine or cervical lymph nodes.

Chemotherapy: Treatment of disease by means of chemical substances or drugs.

Chondroblast: A cell of growing cartilage tissue that is rapidly dividing and undifferentiated; intermediate in radiosensitivity.

Chondrocyte: A nondividing, differentiated mature cartilage cell; highly radioresistant.

Chromatid: Each of the two strands formed by duplication of a chromosome that becomes visible during prophase of mitosis or meiosis.

Chromatid Aberration: Any deviation from the normal structure of chromatids produced when irradiation occurs *after* DNA synthesis, affecting only one chromatid of a pair.

Chromosome: Units of genetic material responsible for directing cytoplasmic activity and transmitting hereditary information in cells.

Chromosome Aberration: Any deviation from the normal structure of chromosomes produced when irradiation occurs *prior* to DNA synthesis; both chromatids exhibit the change in structure.

Chromosome Ring: A chromosome with ends joined to form a circular structure.

Cirrhosis: Fibrosis, especially of the liver, with hardening caused by excessive formation of connective tissue followed by contraction.

CNS (Central Nervous System): Including brain and spinal cord.

Cobalt-60: A heavy radioactive isotope of cobalt of mass number 60 with a half-life of 5.3 years; emits β-particles and γ-rays; used in radiation therapy in teletherapy and brachytherapy treatment.

CODE (Genetic): The system whereby particular combinations of three adjacent nucleotides in a DNA molecule control the insertion of specific amino acids in a protein molecule.

Collimator: A device of high absorption coefficient material used in diagnostic and therapy units to restrict and confine the x-ray beam to a given area; in nuclear medicine, it restricts the detection of emitted radiations from a given area of interest.

Congenital: Existing at or dating from birth: acquired during in utero development and not through heredity.

Cristae: Shelf-like structures inside the mitochondria; site of specific enzymes necessary for the production of energy.

Critical Organ: The dose-limiting factor in a radiopharmaceutical procedure; that organ which, although it may not accumulate the greatest percentage of a radiopharmaceutical, dictates the amount of a radiopharmaceutical that can be administered.

Curie: A unit of radioactivity in which the number of disintegrations per second equals 3.74×10^{10}; abbreviated Ci.

Cysteamine: A radioprotectant amino acid containing sulfhydryl group.

Cystine: An amino acid found in most proteins; contains a sulfhydryl group; an effective radioprotectant.

Cytokinesis: Division of the cytoplasm during telophase of mitosis or meiosis.

Cytoplasm: The protoplasm outside the nucleus which is the site of all metabolic functions in the cell.

Cytosine: One of two pyrimidines found in both RNA and DNA.

Dermis: The sensitive vascular layer of the skin beneath the epidermis.

Desquamation: The denudation or peeling of the skin surface.

Dicentric: A chromosome having two centromeres.

Differentiating Intermitotic Cells (DIM): Cells produced by division of VIM (vegetative intermitotic cells) and, although actively mitotic, are more differentiated than VIM cells; these cells are less sensitive (or more resistant) to radiation than are VIM cells; e.g., intermediate and Type B spermatogonia.

Differentiation: The sum of the processes whereby undifferentiated cells become specialized functionally and/or morphologically (structurally).

DIM: Differentiating intermitotic cells.

Diploid: The number of chromosomes in somatic cells ($2n$).

Direct Action: The interaction and absorption of an ionizing particle by a biologic macromolecule such as DNA, RNA, protein, enzyme, etc. in the cell.

Disaccharide: A carbohydrate formed from two monosaccharides, e.g., table sugar.

Division Delay: Synonym for mitotic delay.

DNA (Deoxyribonucleic Acid): Double-stranded, helical structure in nucleus that contains the genetic material of the cell; composed of nitrogenous bases, 5-carbon sugars and phosphoric acid.

DNA Synthesis: The process that doubles the amount of original DNA in such a manner that the newly formed DNA is identical to the original molecule.

D_0: An expression of radiosensitivity; graphically derived from the exponential portion of the cell survival curve.

Dose Modifying Compounds: Radioprotectant compounds that act by reducing the effective dose of radiation to cells.

Dose Rate: Rate at which radiation is delivered.

Dosimetry: The concept and measurement of quantity of radiation, either emitted by various sources or absorbed by body tissues.

Doubling Dose: The unit of measurement for the determination of radiation effects on mutation frequency; defined as that dose of radiation which ultimately doubles the number of spontaneous mutations that arise in one generation, generally 50 R in humans.

Doubling Time: The time in which a tumor doubles in volume.

D$_q$: Quasi-threshold dose, the width of the shoulder region of a cell survival curve.

DRF: Dose reduction factor, the ratio of the radiation dose necessary to produce a given effect in the presence of a protecting compound to the radiation dose necessary to produce the same effect in the absence of the same compound.

Edema: An abnormal accumulation of serous fluid in connective tissue or in a serous cavity.

EKG (Electrocardiogram): A graphic record of the heart's action.

Elective Booking: A procedure used in diagnostic radiology and nuclear medicine to identify the potentially pregnant patient.

Electromagnetic Radiation: Radio waves, x-rays, γ-rays, etc. radiation having no mass or charge, often spoken of in terms of photon or quanta—a small packet of energy, e.g., visible light, x-rays and γ-rays.

Embryo: In humans, the developing organism from conception to the end of the sixth week of gestation.

Enzyme: A protein that acts as a catalyst, inducing chemical changes in other substances, itself remaining apparently unchanged in the process.

Epidemiology: A science that deals with the incidence, distribution and control of disease in a population.

Epidermis: Outer nonsensitive and nonvascular layer of the skin overlying the dermis.

Epilation: Hair loss; synonym for alopecia.

Equatorial Plate: The imaginary line in the center of the cell equidistant from both poles along which the chromosomes align during metaphase of mitosis and meiosis.

ER (Endoplasmic Reticulum): A double-membraned cell organelle composed of an irregular network of branching and connecting tubules which functions in the synthesis of proteins and other substances.

Ergastoplasm: Synonym for endoplasmic reticulum.

Erythema: Redness of the skin.

Erythroblast: The first specifically identifiable precursor cell in red blood cell formation; located in red bone marrow.

Erythrocyte: Red blood cell; highly radioresistant.

Esophagitis: Inflammation of the mucous membranes of the esophagus.

Etiology: Cause or origin of a disease or abnormal condition.

Exencephaly: A condition in which the skull is defective, the brain being exposed or extruding; brain hernia.

External Beam Therapy: See teletherapy.

Fetus: In humans, the unborn developing human from 6-week post-conception to birth.

Fibroblasts: Elongated cells that comprise connective tissue; intermediate radiosensitivity.

Fibrocyte: Cells of mature connective tissue; relatively radioresistant.

Fibrosis: The formation of fibrous tissue, usually as a reparative or reactive process.

5 BUdR: A halogenated pyrimidine; a true radiosensitizing drug.

5 FU: Fluoracil—a halogenated pyrimidine; not a true radiosensitizing drug.

5 IUdR: 5-iodouridine; a true radiosensitizing drug.

Fixed Postmitotic Cells: Cells that do not divide and are highly differentiated both morphologically and functionally; most radioresistant of all cells.

Fluoroscope: An apparatus for visualizing the shadows of x-rays which, after passing through the body, are projected on a fluorescent screen.

FPM: Fixed postmitotic cells.

Fractionation: The extention of the total radiation dose in radiation therapy over a period of time, ordinarily days or weeks, in order to minimize untoward radiation effects on normal contiguous tissue.

Free Radical: A radical in its (usually transient) uncombined state; an atom or atom group carrying an unpaired electron and no charge; an atom containing a single unpaired electron, rendering it highly reactive.

Gamete: Germ cells in sexually reproducing plants and animals; female—oocytes and male—spermatozoa.

γ-Rays: Electromagnetic radiations originating from atomic nuclei, produced when an unstable atomic nucleus releases energy to gain stability.

Gastrointestinal: Relating to both stomach and intestine.

Gene: Located on a chromosome, the unit of genetic material that is responsible for directing cytoplasmic activity and transmitting hereditary information in the cell.

Germ Cell: Cells in sexually reproducing plants and animals with a haploid number of chromosomes.

Gestation: The carrying of the young in the uterus: pregnancy.

Gonad: An organ that produces sex cells; the testis of a male or the ovary of a female.

G_1—Gap 1: A period between telophase and the beginning of DNA synthesis when DNA is not replicating.

Granulocyte: A mature leukocyte containing cytoplasmic granules.

Growth Fraction: The fraction of the total tumor cell population that is dividing, responsible for tumor growth.

GSD (Genetic Significant Dose): An average figure calculated from the actual gonadal doses received by the exposed population; evaluates the genetic impact of medical/dental x-ray doses on the whole population.

G_2—Gap 2: Following DNA replication and prior to mitosis.

Guanine: One of two purine bases found in both DNA and RNA.

Halogenated Pyrimidines: Chemical compounds that can replace the base thymine on DNA; effective radiosensitizers including 5-bromodeoxyuridine (5-BUdR) and 5-iododeoxyuridine (5-IUdR).

HeLa Cells: Cells originally derived from a human carcinoma of the cervix; now grown in tissue culture.

Helix: A spiraled structure.

Hemoglobin: The red protein of erythrocytes that carries O_2.

Hemopoietic System: Pertaining to the system that deals with the formation of blood cells including the bone marrow, circulating blood, lymph nodes, spleen and thymus.

Hemorrhage: Bleeding; a flow of blood, especially if profuse.

Hepatic: Relating to the liver.

Hepatitis: Inflammation of the liver.

Hodgkin's Disease: A malignant disease of lymph nodes, characterized by enlargement of the nodes, often cervical in onset and then generalized, sometimes with enlargement of the spleen and liver.

Homologous: Having the same relative position, value or structure; applied to chromosomes; two chromosomes with identical sequence and type of genes.

Hydrocephaly: Water on the brain.

Hydroxyurea (HU): A drug that synchronizes a population of cells; sometimes used in combination with radiation therapy.

Hyperbaric Oxygen (HPO): Pure O_2 at 3 atmospheres of pressure, tried in radiation therapy to increase available O_2 to tumor cells and overcome the oxygen effect.

Hyperthyroidism: Excessive functional activity of the thyroid gland resulting in a condition marked by increased metabolic rate and enlargement of the thyroid gland.

Hypoplasia: Loss of cells, tissue or organ atrophy due to destruction of some of the elements and not merely to their general reduction in size.

Hypoxia: Decreased amount of oxygen; oxygen deficiency, less than physiologically normal amount.

Hypoxic Cell Sensitizers: Drugs that selectively sensitize hypoxic cells, one example is the nitrofurans, such as NDPP-*p*-nitro, 3-dimethylaminopropiophenone hydrochloride.

Immune System: Those cells (e.g., lymphocytes and other white cells) and organs, (spleen, lymph nodes and thymus) which provide natural defense against disease to the body.

Immunotherapy: Treatment directed at the immune system.

In Utero: Within the womb; not yet born.

In Vitro: In glassware or, in an artificial environment, outside the living body; opposite of in vivo—in a living system.

In Vivo: In the living body; referring to vital chemical processes etc. as distinguished from those occurring in the test tube (in vitro).

Indirect Action: The absorption of ionizing radiation by the medium in which cell organelles are suspended, primarily water.

Inflammation: A local response to cellular injury marked by capillary dilation, leukocytic infiltration, redness, heat, pain and swelling; serves as a mechanism initiating the elimination of toxic agents and of damaged tissue.

Inorganic: Not organic; not relating to living organisms; in chemistry, referring to compounds not containing carbon, e.g., mineral salts (Na + K).

Insulin: A well-known protein (polypeptide) produced by the pancreas involved in the regulation of metabolism, especially carbohydrate metabolism.

Integral Dose: Incorporates volume, in grams, of tissue irradiated; defined as gram-rads.

Interphase: The nondividing or intermitotic period between two successive divisions of a cell.

Interphase Death: Cell death before entering mitosis.

Interstitial Therapy: Placement of sealed radioactive sources directly in a malignant tumor.

Intracavitary Therapy: Placement of sealed radioactive sources within a body cavity close to the tumor; the sources are held in place by applicators, e.g., in the treatment of carcinoma of the cervix.

Inversions: A turning inward, upside down or in any direction contrary to the existing one; a type of chromosome aberration.

Ion: An atom or group of atoms that carries a positive or negative charge as a result of having lost or gained one or more electrons.

Irradiation: Subjection to radiation.

Isoeffect Curve: Similar curve, a line formed from plotting total dose and overall treatment time on a double logarithmic plot; relates treatment schedule to clinical results.

KeV/μ (micron): The unit of expression for LET equal to the energy deposited per unit distance of path traveled by a charged particle.

Leiberkuhn, Crypts of: Named after the German anatomist. Nests of cells at the base of the villi in the intestine; a rapidly dividing, undifferentiated stem cell population; relatively radiosensitive.

Lesion: An abnormal change in structure of an organ or part due to injury or disease.

LET: Linear energy transfer.

Leukemia: An acute or chronic disease of unknown cause in humans and other warm-blooded animals characterized by an abnormal increase in the number of leukocytes (white blood cells) in tissue and blood.

Leukemogenic: Leukemia-causing.

Leukocyte: Any of the white or colorless nucleated cells that occur in blood.

Leukopenia: Pronounced reduction in the number of white blood cells in the circulating blood.

Linear Accelerator: A device in which electrons are accelerated striking a target or window, generating x-rays greater than 4 mv; teletherapy treatment unit used in radiation therapy.

Linear Energy Transfer: The rate at which energy is deposited as a charged particle travels through matter.

Lipid: A class of organic compounds that, with proteins and carbohydrates, constitute the principal structural components of living cells. Functions include energy storage, protection of the body against cold and assistance in digestive processes.

Lumbar: Relating to the loins; the part of the back and sides between the ribs and the pelvis.

Lymphocytes: White blood cells formed in lymphoid tissue throughout the body, e.g., lymph nodes, spleen, thymus, tonsils, Peyer's patches and sometimes bone marrow; highly radiosensitive.

Lymphoid: Resembling lymph (a pale coagulable fluid that consists of a liquid portion resembling blood plasma containing white blood cells).

Lymphoma: A general term that includes various, abnormally proliferative neoplastic diseases of the lymphoid tissues, e.g., lymphosarcoma and Hodgkin's disease; ordinarily termed malignant lymphoma.

Lyse: To break up, to disintegrate, e.g., as applied to cells.

Lysosome: A single-membraned cell organelle that contains enzymes capable of breaking down proteins, DNA and some carbohydrates.

M-Mitosis: Physical division of the cell; consists of prophase, metophase, anaphase, and telophase; a very radiosensitive phase of the cell cycle.

Macromolecule: Polymers, notably proteins, nucleic acids and polysaccharides.

Malignant: Resistant to treatment, occurring in severe form and frequently fatal; tending to become worse; in the case of a neoplasm, having properties of invasion, metastases and uncontrollable growth.

Mammal: Any of a class (mammalia) of higher vertebrates comprising humans and all other animals that nourish their young with milk secreted by mammary glands and have the skin usually more or less covered with hair.

Mastectomy: Amputation of the breast.

Matrix: The ground cytoplasm, or medium, in which the cell organelles are suspended.

Maturation Depletion: Process of depletion of mature sperm by depopulation of spermatogonia.

Maximum Permissible Dose: The allowable radiation dose at which there is assumed to be relatively small biologic risk to occupationally exposed persons and to the general population; the present MPD for total body exposure of occupationally exposed persons is 5 rem/year after the age 18.

Megakaryocytes: Precursor cells for blood platelets; the least radiosensitive of bone marrow stem cells.

Meiosis: Process of cell division in germ cells, consisting of two cellular divisions but only one DNA replication, resulting in formation of four gametocytes each containing half the number of chromosomes found in somatic cells.

Melanoma: A malignant neoplasm derived from cells that are capable of forming melanin.

Meson: An unstable nuclear particle first observed in cosmic rays; has a mass between that of the electron and the proton; either charged (positively or negatively) or neutral; investigational radiation therapy modality.

Metabolism: The sum of the processes in the building up (anabolism) and destruction (catabolism) of protoplasm incidental to the cell life.

Metaphase: The stage of mitosis or meiosis in which the chromosomes become aligned on the equatorial plate of the cell with the centromeres mutually repelling each other.

Metastasis: The appearance of neoplasms in parts of the body remote from the site of the primary tumor.

Methotrexate (MTX): A drug used to synchronize a population of cells, sometimes used in combination with radiation therapy.

Microcephaly: Abnormal smallness of the head (brain).

Microcurie: One millionth of a curie; 3.7×10^4 disintegrations per second; abbreviated μCi.

Microphthalmia: Small size of one or both eyeballs.

Millicurie: One-thousandth of a curie; 3.7×10^7 disintegrations per second; abbreviated mCi.

Mitochondria: Organelles of the cell cytoplasm consisting of two sets of membranes, a smooth continuous outer coat and a convoluted inner membrane arranged in folds that form shaft-like structures called cristae; the powerhouses of the cell, producing energy for cellular functions.

Mitosis: The process of cellular reproduction in somatic cells whereby one "parent cell" divides to form two "daughter cells" with the same chromosome number and DNA content as the original "parent cell"; consists of four phases—prophase, metaphase, anaphase and telophase.

Mitotic Delay: A cellular response to irradiation; cells are delayed from entering mitosis for a varying period of time, dependent on dose.

Mitotic Index: The ratio of the number of cells in mitosis at any one time to the total number of cells in the population.

Molecule: The smallest possible quantity of a substance, composed of two or more similar or dissimilar atoms that exist independently and still retain the chemical properties of the substance of which it forms a part, e.g., O_2, H_2O and H_2.

Mongoloid: A child born with a congenital idiocy of unknown cause having slanting eyes, a broad short skull and broad hands with short fingers.

Monomer: Simple molecular units that are individually stable and have specific chemical characteristics and properties.

Monosaccharide: A carbohydrate not decomposable to simple sugars; e.g., glucose and fructose.

Morphology: Study of the structure or form of biologic materials.

MPD: See Maximum Permissible Dose.

Mucositis: Inflammation of the mucous membranes of the oral cavity.

Multidisciplinary Approach: A combination of cancer treatment modalities, e.g., radiation therapy, surgery, chemotherapy and immunotherapy.

Multipotential Connective Tissue Cells: Cells that divide irregularly and are more differentiated than either VIM (vegetative intermitotic cells) or DIM (differentiating intermitotic cells) cells; intermediate radiosensitivity, e.g., endothelial cells.

Mutagen: Any agent that causes the production of a mutation, e.g., virus, drugs and radiation.

Mutation: Alteration in the sequence of base pairs on the DNA molecule or in the amount or volume of DNA.

Mutation Frequency: The number of spontaneous or induced mutations that arise per generation.

Myelitis: Inflammation of the spinal cord.

Myelocytes: Precursor cells in the bone marrow to circulating white cells; relatively radiosensitive.

Myeloid: (1) Of or relating to the spinal cord. (2) Of or relating to or resembling bone marrow.

n (Extrapolation Number): One of three graphic parameters used to define the cell survival curve; determined by extrapolating the exponential portion of the cell survival curve to its intersection with the Y axis.

n Number (Haploid Number): Refers to the number of chromosomes in germ cells.

NCRP: National Commission on Radiation Protection and Measurement.

Necrosis: The pathologic death of one or more cells or of a portion of tissue or organ, resulting from irreversible damage.

Neonatal: Relating to the period immediately succeeding birth and continuing through the first month of life.

Neoplasm: A new growth, e.g., a tumor, not necessarily malignant.

Nephritis: Acute or chronic inflammation of the kidneys.

Neuroblast: An embryonic nerve cell.

Neutron: An uncharged particle in the nucleus of an atom.

Nominal Standard Dose (NSD): A unit incorporating total tumor dose, number of fractions and overall treatment time, expressed in the formula $D = (NSD)T^{0.11}N^{0.24}$, where D = total dose, N = number of fractions and T = overall treatment time.

Nondivision Death: See interphase death.

Nonmitotic Death: See interphase death.

NSD: See Nominal Standard Dose.

Nuclear Medicine: The clinical field of study concerned with the diagnostic and therapeutic use of radionuclides, excluding the therapeutic use of sealed radiation sources.

Nucleic Acid: Major class of organic compounds in the cell composed of a sugar or derivative of a sugar, phosphoric acid and a base, i.e., RNA and DNA.

Nucleolus: Mass of stainable material in the cell nucleus that houses nuclear RNA.

Nucleotide: The three components of DNA taken together including a nitrogenous base, a 5-carbon sugar and phosphoric acid.

Nucleus: Center; applied to cells, a portion of protoplasm containing the genetic material separated from the cytoplasm by a membrane.

Occlude: To close or bring together.

OER (Oxygen Enhancement Ratio): Ratio of doses to produce a given biologic response in the absence of oxygen to the dose to produce the same response in the presence of oxygen.

Oocyte: The immature ovum.

Organ: A differentiated structure consisting of cells and tissues and performing some specific function such as respiration, secretion or digestion.

Organelle: Specialized structures in a cell performing specific functions.

Organic: Of or relating to or containing carbon compounds.

Organogenesis: The formation of organs during development; in humans the period from the second to sixth week postconception.

Orthovoltage: Medium voltage of 250 kv, referred to in radiation therapy as deep therapy.

Osmotic Pressure: The force under which a solvent moves from a solution of lower solute concentration to a solution of higher solute concentration when these solutions are separated by a selectively permeable membrane.

Osteoblast: Bone-forming cells; moderately radiosensitive.

Osteoclast: Bone-resorbing cells; moderately radiosensitive.

Osteosarcoma: Malignant neoplasm of bone.

Ovarian Follicle: One of the sac-like enclosures in the ovary containing an ovum.

Overshoot, Mitotic: An increased number of cells in mitosis.

Oxygen Effect: Specific name given to the response of cells to radiation in the presence of oxygen.

Oxygen Enhancement Ratio: See OER.

Oxygen Tension: The pressure excreted by oxygen proportional to the percentage of oxygen molecules present in the total volume of blood.

Palliative: Relief; reduce in severity; often used in relation to a method of treatment of a disease.

Pancarditis: Inflammation of all the structures of the heart, including endocardium, myocardium and epicardium.

Pancreas: A large compound gland of vertebrates that secrete digestive enzymes and the hormone insulin.

Pancytopenia: Pronounced reduction in the number of erythrocytes, all types of white blood cells and blood platelets in the circulating blood.

Papain: A proteolytic enzyme used as a protein digestant and as a meat tenderizer.

Parenchyma: The distinguishing of specific cells of a gland or organ, contained in and supported by the connective tissue framework or stroma.

Particle: A type of ionizing radiation that has mass and charge.

Peptide Bond: A chemical bond joining two amino acids.

Pericarditis: Inflammation of the pericardium—the membrane covering the heart.

Petechial Hemorrhages: Minute hemorrhagic spots, of pinpoint to pinhead size, in blood vessel walls.

Physical Half-Life: The time required for one-half the number of atoms of a specific radionuclide to undergo disintegration.

Physiology: Study of the normal vital processes of living animal and vegetable organisms.

Pion: Pi-meson.

Pneumonitis: Inflammation of the lungs.

Polycythemia Vera: A malignant disease of red blood cells; actually an increased number of RBC's.

Polymer: Two or more monomers joined by polymerization to form a chain consisting essentially of repeating structural units.

Polymerization: The process by which polymers are joined.

Polysaccharide: A carbohydrate that can be decomposed into two or more molecules of monosaccharides.

Portal: Opening or window; in radiation therapy, that area of the body being treated.

Potentiate: To enhance; in radiation, to enhance the response of cells, tissues, organs, etc. by addition of some substance, e.g., oxygen.

Precursor: Anything that precedes another or from which another is derived; forerunner.

Primary Repair: Replacement of damaged cells in the organ by the same cell type present before radiation; also called *regeneration*.

Procreate: To produce offspring; *generation*.

Prodromal Stage: Initial; in radiation, that stage of response to an acute total body dose of radiation characterized by nausea, vomiting and diarrhea (often referred to as N-V-D syndrome).

Prophase: The first stage of mitosis or meiosis consisting of linear contraction and increasing thickness of the chromosomes accompanied by division of the centriole.

Protein: One of the four major classes of cellular organic compounds; macromolecules consisting of long sequences of α-amino acids in peptide linkage; constitute approximately 15% of cell content and are the most plentiful carbon-containing compounds in the cell.

Protoplasm: The colloidal complex of organic and inorganic substances and water that constitutes the living cell.

Puck, T. T., and Marcus, T. I.: Two investigators who conducted a classic radiobiologic experiment and constructed cell survival curve.

Purine: One of two categories of nitrogenous bases found in DNA; includes the bases adenine and guanine.

Pyelogram: A radiogram of the renal pelvis and uterer.

Pyrimidine: One of two categories of nitrogenous bases found in DNA, includes the bases thymine and cytosine.

Quasi-Threshold Dose (D_q): Defines the width of the shoulder region of the cell-survival curve and is the dose at which point the curve becomes exponential.

Rad: A unit of absorbed dose, equal to 100 ergs/gm.

Radiation Syndrome: A group of signs and symptoms that occur together, characterizing the system most affected by total body exposure to ionizing radiation.

Radiation Tolerance: The limit of radiation exposure a normal tissue can receive and still remain functional.

Radical (Noun): A group of elements or atoms usually passing intact from one compound to another but usually incapable of prolonged existence in a free state.

Radiobiology: Study of the effects of ionizing radiation on living things.

Radionuclide: Unstable form (or forms) of a given element all having the same number of protons but a varying number of neutrons.

Radiopharmaceutical: A chemical or drug to which radionuclide has been added.

Radioprotectants: Chemicals or drugs that diminish the response of cells to radiations.

Radiosensitizers: Chemicals and drugs that enhance radiation response of cells.

Radiotherapy: The treatment of disease with ionizing radiation.

Radium: A radioactive, shiny white, metallic element that resembles barium chemically, occurs in combination in minute quantities in minerals (e.g., pitchblende), emits α-particles and γ-rays to form radon and is used chiefly in luminous materials and in the treatment of cancer.

RBC: Erythrocyte or red blood cell; relatively radioresistant.

RBE: Relative biologic effect:

$$RBE = \frac{\text{Dose in Rads from 250 keV x-ray}}{\text{Dose in Rads from another radiation}}$$

to produce the same biologic response

Regeneration: The replacement of damaged cells in the organ by the same cell type present before radiation; also called primary repair.

Relative Biologic Effect: A term relating the ability of radiations with different LET ranges to produce a specific biologic response; the comparison of a dose of test radiation to a dose of 250 KeV x-ray both producing the same biologic response.

REM: A unit of dose equivalent that is numerically equal to the dose in rads times quality factor times modifying factor.

Replication: Repeated formation of the same molecule, as of DNA.

Reproductive Failure: The inability of the cell to undergo repeated divisions after irradiation.

Resectable: Amenable to surgical removal.

Resolve: To find an answer to; to clear up.

Respiratory System: A system of organs serving the function of respiration consisting of the nose, pharynx, trachea and lungs.

Restitution: Restoration, as in chromosome restitution.

RET (Rad Equivalent Therapy): Unit of expression of NSD.

Reverting Postmitotic Cells: Cells that do not normally undergo mitosis but retain the capability of division under specific circumstances; relatively radioresistant.

Ribonucleic Acid: One of two general types of nucleic acid that functions primarily in protein synthesis and has D-ribose as its sugar constituent and adenine, guanine, cytosine and uracil as bases and is found both in the nucleus and cytoplasm of cells.

Ribosome: A protoplasmic organelle containing ribonucleic acid; necessary for protein synthesis.

RNA: Ribonucleic acid.

RPM: See reverting postmitotic cells.

"S" Period: Synthesis; replication; phase of cell after G_1 and prior to G_2 when DNA is duplicated.

Sarcoma: A malignant neoplasm arising from relatively nonrenewing tissue of mesodermal origin (as connective tissue, bone, cartilage or striated muscle).

Sclerosis: Pathologic hardening of tissue especially from overgrowth of fibrous tissue or increase in interstitial tissue.

Scoliosis: Lateral curvature of the spine.

Sebaceous Gland: One of a large number of glands in the dermis that usually open into the hair follicles and secrete an oily, semifluid substance.

Secondary Repair: The replacement of the original radiation damaged cells by a different cell type—usually one which forms connective tissue resulting in a scar; also called fibrosis.

Sensitivity, Conditional: The modification of cellular radiation response by external factors.

Sensitivity, Inherent: The response of the cell to radiation due to characteristics specific to the cell (mitotic activity and differentiation).

Skin Erythema Dose (SED): That acute dose of radiation (1000 rads) which causes skin erythema.

Somatic Cells: All cells of the body other than germ cells.

Spermatozoa: The mature, differentiated, functional cells of the testes.

Spina Bifida: A defect in the spinal column in which there is absence of the vertebral arches and through which the spinal membranes, with or without spinal cord tissue, may protrude.

"Split-Course" Treatment Schedule: Administering the total therapeutic dose in two segments or courses, separated by a variable rest interval during which treatment is not given.

Spontaneous Mutation: A naturally occurring alteration in the structure, volume or amount of DNA.

Stickiness, Chromosome: Clumping of chromosomes due to irradiation.

Strandqvist, M.: Did classic study in 1944 relating various fractionation schedules to the cure of skin cancer and normal tissue tolerance.

Stricture: A circumscribed narrowing or stenosis (closing) of a tubular structure, e.g., of the esophagus.

Stroma: The supporting framework, usually of connective tissue, of an organ, gland or other structure; distinguished from the parenchyma.

Sulfhydryl: Sulfur and hydrogen bound together, designated SH.

Supervoltage: Megavoltage; greater than 1000 volts.

Synchronous: Occurring simultaneously.

Syndrome: A group of signs and symptoms that occur together and characterize a particular abnormality or disease.

Synthesis: (1) The formation of compounds by the union of simpler compounds or elements. (2) A building up; a putting together.

Systemic: Relating to a system or to the entire organism.

Systems: Multicellular—a group of organs made up of millions of cells that together perform one or more vital functions, e.g., digestion, respiration or circulation.

Target Organ: The organ of interest in a radiopharmaceutical or radionuclide procedure.

TD 5/5: That dose which, when given to a population of patients, results in the minimum (5%) severe complication rate within 5 years posttreatment.

TD 50/5: That dose which, when given to a population of patients, results in the maximum (50%) severe complication rate in 5 years.

Technetium 99^m: ^{99m}Tc; a metastable radionuclide formed by the decay of molybdenum, decays by gamma emission and has a half-life (physical) of 6 hours, widely used in nuclear medicine.

Telangiectasis: Dilation of previously existing small or terminal vessels of an organ.

Teletherapy: Radiotherapy treatment "at a distance," generally refers to treatment administered from outside the body.

Telophase: The final stage of mitosis or meiosis beginning when migration of chromosomes to the poles of the cell is complete.

10-Day Rule: A rule suggested by the National Commission on Radiation Protection and Measurement recommending that all nonemergency diagnostic procedures involving the pelvis of women of childbearing age be conducted during the first 10 days of the menstrual cycle when the probability of pregnancy is very small.

Thoracic: Relating to the thorax or chest.

Threshold: The point where a stimulus begins to produce a sensation; the lower limit of perception of a stimulus.

Thrombosis: The formation of a blood clot.

Thymine: A pyrimidine nitrogenous base of DNA only.

Tissue Culture: Growing of animal and human cells in a bottle or tube by providing nutrients.

Tolerance Dose: A concept that expresses the minimal and maximal injuries acceptable for different organs and the dose at which they occur.

Translocation: The transposition of two segments between nonhomolgous chromosomes as a result of abnormal breakage.

Transverse Myelitis: Inflammation involving the entire thickness of the spinal cord.

Trauma: An injury (as a wound) to living tissue caused by an extrinsic agent.

Treatment Field: In radiation therapy, that area of the body being treated.

2N Number (Diploid Number): The number of chromosomes in somatic cells.

250 KeV X-Ray: Orthovoltage radiation.

Vasculature: Relating to or containing blood vessels.

Vegetative Intermitotic Cells: A group of rapidly dividing undifferentiated cells that have a short lifetime; most radiosensitive.

Villi: Finger-like projections in the lining of the intestine.

Vim: See vegetative intermitotic cells.

Virus: Any of a large group of submicroscopic agents consisting of nucleic acids and a protein coat that are capable of growth and multiplication only in living cells, causing various diseases.

X-Rays: Electromagnetic radiations originating from the orbital electrons of an atom.

Zygote: The diploid cell resulting from union of a sperm and an ovum.

Index